# Mastering Self-Sufficiency & Homesteading

A Beginner's Guide to Modern & Traditional Skills That Will Improve Your First Homestead or Current Backyard Homesteading, Everything from Planting to Preservation

Sylvia Soilworth

# Contents

# Chapter 1

# Getting Started

Embarking on your homesteading journey begins with careful planning and preparation. This chapter will guide you through the essential steps of getting started, from setting clear goals to designing an efficient layout for your homestead.

## Planning Your Homestead

Planning is the cornerstone of a successful homestead. Without a solid plan, you may find yourself overwhelmed and unsure of where to start. Here are the key elements to consider when planning your homestead:

## Vision and Purpose

Begin by defining your vision for your homestead. What are your long-term goals? Do you want to become entirely self-sufficient, or are you looking to supplement your current lifestyle with homegrown produce and homemade goods? Understanding your purpose will help guide your decisions and keep you motivated.

## Research and Education

Educate yourself about homesteading practices, local regulations, and climate considerations. Books, online resources, and local extension offices can provide valuable information. Attend workshops, join homesteading groups, and connect with experienced homesteaders to learn from their experiences.

## Budgeting

Determine your budget for starting and maintaining your homestead. Consider the cost of land, infrastructure, tools, seeds, livestock, and other necessary supplies. Create a financial plan that outlines initial investments and ongoing expenses. Look for ways to minimize costs through DIY projects, bartering, and repurposing materials.

## Timeline

Develop a timeline for achieving your goals. Break down your plan into manageable phases, and set realistic deadlines for each phase. This will help you stay organized and focused, ensuring steady progress.

## Setting Goals and Objectives

Setting clear, achievable goals is essential for the success of your homestead. Goals provide direction and motivation, helping you stay on track and measure your progress. Here's how to set effective goals and objectives:

## Short-Term and Long-Term Goals

Distinguish between short-term and long-term goals. Short-term goals are those you can achieve within a few weeks or months, such as planting your first garden or building a chicken coop. Long-term goals might include establishing a renewable energy system or becoming fully self-sufficient.

## Specific, Measurable, Achievable, Relevant, and Time-Bound (SMART) Goals

Use the SMART criteria to set your goals:

- **Specific:** Clearly define what you want to achieve. For example, "Grow enough vegetables to supply our family with fresh produce year-round."

- **Measurable:** Ensure your goals can be quantified or measured. For example, "Produce 100 pounds of tomatoes by the end of the summer."

- **Achievable:** Set realistic goals that are within your capabilities and resources. For example, "Raise five chickens for eggs and meat."

- **Relevant:** Align your goals with your overall vision and purpose. For example, "Install a rainwater harvesting system to reduce our reliance on municipal water."

- **Time-Bound:** Set a deadline for achieving your goals. For example, "Build a greenhouse by the start of the next growing season."

## Prioritizing Goals

Prioritize your goals based on their importance and feasibility. Focus on the most critical tasks first, such as securing water and shelter, before moving on to more ambitious projects. This approach ensures that your basic needs are met and provides a solid foundation for future growth.

## Assessing Your Space and Resources

A thorough assessment of your space and resources is crucial for effective planning. Understanding the strengths and limitations of your property will help you make informed decisions and optimize your efforts.

## Evaluating Your Land

Consider the following factors when evaluating your land:

- **Size and Shape:** Determine the total area available for your homestead and its shape. This will influence your layout and the types of activities you can undertake.

- **Topography:** Assess the slope and elevation of your land. Flat areas are ideal for gardening, while sloped areas can be used for terracing or grazing livestock.

- **Soil Quality:** Test your soil to determine its composition, pH, and nutrient levels. This information will help you choose appropriate crops and soil amendments.

- **Water Sources:** Identify natural water sources such as ponds, streams, or wells. Consider how you will access and distribute water for irrigation, livestock, and household use.

- **Climate and Microclimates:** Understand the climate of your region, including average temperatures, rainfall, and frost dates. Identify microclimates on your property that may be more suitable for certain plants or activities.

## Existing Infrastructure

Take stock of any existing infrastructure on your property, such as buildings, fences, or utilities. Determine how these structures can be utilized or modified to support your homesteading activities. Consider the condition and potential uses of barns, sheds, greenhouses, and other outbuildings.

## Natural Resources

Identify and assess the natural resources available on your property. These might include:

- **Woodlands:** Trees can provide firewood, lumber, and other materials. They also offer habitat for wildlife and can be managed for sustainable forestry.

- **Pastures and Meadows:** Open areas can be used for grazing livestock, growing hay, or establishing orchards.

- **Wildlife:** Recognize the presence of beneficial wildlife, such as pollinators and predators, as well as potential pests.

## Designing Your Farm Layout

A well-designed farm layout maximizes efficiency, productivity, and sustainability. Thoughtful planning can reduce labor, conserve resources, and enhance the overall functionality of your homestead.

## Zoning and Permaculture Principles

Consider using permaculture principles to design your farm layout. Permaculture emphasizes working with nature to create sustainable and self-sufficient systems. One key concept is zoning, which involves organizing your homestead into zones based on the frequency of use and the needs of different elements.

- **Zone 0:** Your home and immediate surroundings. This area includes your living space, kitchen garden, and frequently used outdoor areas.

- **Zone 1:** Areas visited daily, such as vegetable gardens, chicken coops, and compost bins.

- **Zone 2:** Areas visited several times a week, such as larger livestock enclosures, perennial gardens, and orchards.

- **Zone 3:** Areas visited weekly or monthly, such as fields, pastures, and woodlots.

- **Zone 4:** Semi-wild areas used occasionally for foraging, hunting, or timber production.

- **Zone 5:** Wild areas left undisturbed to support biodiversity and ecosystem health.

## Layout Considerations

When designing your layout, consider the following factors:

- **Accessibility:** Arrange elements to minimize travel time and effort. Place frequently used areas close to your home, and ensure paths are clear and easy to navigate.

- **Water Management:** Plan for efficient water use and distribution. Consider the natural flow of water on your property and design systems to capture, store, and distribute water where needed.

- **Sunlight and Shade:** Position gardens and structures to take advantage of sunlight and shade. Orient rows and beds to maximize sun exposure for crops and provide shade for animals.

- **Wind and Weather:** Use natural features and structures to protect against wind and extreme weather. Plant windbreaks, build shelters, and consider microclimates when positioning elements.

- **Soil and Drainage:** Design your layout to prevent soil erosion and improve drainage. Use contour planting, swales, and terraces to manage water flow and maintain soil health.

## Mapping and Planning

Create a detailed map of your property to visualize your layout. Use graph paper or digital tools to draw a scaled representation of your land. Include existing structures, natural features, and planned elements. This map will serve as a valuable reference as you develop and refine your homestead.

## Flexibility and Adaptability

While it's important to have a plan, be prepared to adapt as you learn and grow. Homesteading is an evolving process, and you may need to adjust your layout and practices based on experience and changing conditions. Stay flexible and open to new ideas, and be willing to make changes as needed to optimize your homestead's performance.

## Essential Tools for Homesteading

A well-equipped homestead requires a variety of tools to handle different tasks efficiently. Investing in quality tools can save you time, effort, and money in the long run. Here are some essential tools every homesteader should consider:

## Gardening Tools

- **Shovel:** A sturdy shovel is indispensable for digging, planting, and moving soil and compost.

- **Rake:** Use a garden rake for leveling soil, spreading mulch, and collecting debris.

- **Hoe:** A hoe is useful for weeding, cultivating soil, and creating furrows for planting.

- **Garden Fork:** Ideal for turning compost, aerating soil, and harvesting root crops.

- **Pruners and Shears:** Essential for trimming plants, pruning trees, and harvesting produce.

- **Watering Can or Hose:** Ensure your plants get adequate water with a reliable watering system.

## Building and Maintenance Tools

- **Hammer:** A good quality hammer is necessary for construction projects, repairs, and general maintenance.

- **Screwdrivers:** Both flathead and Phillips screwdrivers are needed for assembling and fixing various items.

- **Wrenches and Pliers:** Useful for tightening and loosening nuts and bolts, as well as gripping and cutting wires.

- **Saw:** A hand saw or a power saw is essential for cutting wood for building projects and repairs.

- **Measuring Tape:** Accurate measurements are crucial for any building or planting project.

- **Level:** Ensures that your structures are straight and stable.

## Livestock and Animal Care Tools

- **Feeders and Waterers:** Ensure your animals have access to food and water with appropriate feeders and waterers.

- **Nesting Boxes:** Provide a safe place for chickens to lay eggs.

- **Fencing Tools:** Tools such as post-hole diggers, wire stretchers, and fencing pliers are essential for building and maintaining fences.

- **Brush and Clippers:** Keep your livestock well-groomed and healthy with brushes and clippers.

## Miscellaneous Tools

- **Wheelbarrow or Garden Cart:** Transport soil, compost, plants, and other materials around your homestead.

- **Buckets and Containers:** Useful for carrying water, feed, and harvested produce.

- **Gloves:** Protect your hands from blisters, cuts, and dirt with durable gloves.

- **Tool Sharpeners:** Keep your tools in good working condition with sharpeners for blades and cutting edges.

## Choosing the Right Materials

Selecting the right materials for your homestead projects is essential for durability, functionality, and sustainability. Here are some key considerations when choosing materials:

## Wood

- **Types of Wood:** Choose wood based on its intended use. Hardwoods like oak and maple are durable and suitable for structural projects, while softwoods like pine are easier to work with and good for indoor projects.

- **Treated vs. Untreated:** Use treated wood for outdoor projects like fencing and raised beds to prevent rot and insect damage. However, be cautious with treated wood near edible plants due to potential chemical leaching.

- **Sustainable Sourcing:** Whenever possible, use locally sourced or reclaimed wood to reduce environmental impact and support local economies.

## Metal

- **Galvanized Steel:** Ideal for outdoor structures like fences and animal shelters due to its rust-resistant properties.

- **Aluminum:** Lightweight and corrosion-resistant, suitable for roofing and certain types of fencing.

- **Recycled Metal:** Consider using recycled metal for sustainability and cost savings.

## Stone and Brick

- **Natural Stone:** Durable and aesthetically pleasing, natural stone is ideal for pathways, retaining walls, and foundations.

- **Brick:** A versatile building material that provides excellent insulation and durability for structures like ovens, walls, and raised beds.

- **Sourcing:** Use locally sourced stone and brick to minimize transportation costs and environmental impact.

## Plastics and Composites

- **High-Density Polyethylene (HDPE):** Durable and resistant to impact, HDPE is suitable for water tanks, pipes, and certain types of fencing.

- **Composite Materials:** Often made from recycled materials, composites can be used for decking, fencing, and raised beds.

- **Environmental Considerations:** While plastics are durable, consider their environmental impact and explore alternatives where possible.

## Building Your Farm Infrastructure

Creating a functional and efficient infrastructure is essential for a successful homestead. Here are the primary components to consider when building your farm infrastructure:

## Shelters and Buildings

- **Tool Shed:** A secure, weatherproof shed to store tools, equipment, and supplies.

- **Greenhouse:** Extend your growing season and protect plants from harsh weather with a greenhouse. Consider size, ventilation, and heating options.

- **Animal Shelters:** Provide appropriate shelters for your livestock, such as coops for chickens, barns for larger animals, and hutches for rabbits. Ensure they are well-ventilated, predator-proof, and easy to clean.

- **Storage Buildings:** Additional storage for hay, feed, and harvested produce. Consider rodent-proofing and climate control for sensitive items.

## Fencing and Gates

- **Perimeter Fencing:** Secure your property with durable fencing to keep animals in and predators out. Choose the right type of fencing based on your livestock and local wildlife.

- **Internal Fencing:** Separate different areas of your homestead, such as gardens, pastures, and animal enclosures. Use movable fencing for rotational grazing and to manage space efficiently.

- **Gates:** Install sturdy gates for easy access to different areas of your homestead. Ensure they are wide enough for equipment and livestock to pass through comfortably.

## Water Systems

- **Irrigation:** Implement efficient irrigation systems like drip irrigation or soaker hoses to conserve water and ensure your plants receive adequate moisture.

- **Rainwater Harvesting:** Collect and store rainwater for irrigation and animal use. Set up gutters, downspouts, and storage tanks to maximize water collection.

- **Wells and Pumps:** If you have a well, ensure it is properly maintained and equipped with a reliable pump. Consider solar-powered pumps for sustainability.

## Composting and Waste Management

- **Compost Bins:** Set up compost bins or piles to recycle organic waste into valuable compost for your garden. Choose a system that suits your space and volume of waste.

- **Manure Management:** Properly manage animal manure to prevent contamination and create nutrient-rich fertilizer. Compost or age manure before using it in your garden.

- **Waste Disposal:** Establish a system for disposing of non-compostable waste. Reduce, reuse, and recycle as much as possible to minimize waste.

Getting started with homesteading requires careful planning and thoughtful consideration of your goals, resources, and layout. By setting clear objectives, assessing your space, and designing an efficient farm layout, you can create a solid foundation for a successful and sustainable homestead. Equipping your homestead with the right tools and materials and building a functional infrastructure are foundational steps toward achieving self-sufficiency. By investing in quality tools, choosing sustainable materials, and thoughtfully designing your infrastructure, you can create an efficient and productive homestead that meets your needs and supports your goals. As you embark on this journey, remember that flexibility and continuous learning are key to adapting and thriving in your homesteading endeavors. Stay flexible and open to new ideas, and be willing to make changes as needed to optimize your homestead's performance.

# Chapter 2

# Soil and Fertility

Soil is the foundation of any successful homestead. It supports plant life, provides nutrients, and helps regulate water. Understanding soil composition and structure, testing and improving soil health, and utilizing organic soil amendments are crucial steps in creating a thriving homestead. This section will delve into these topics in detail.

## Soil Composition

Soil is composed of four main components: minerals, organic matter, water, and air. Each of these components plays a vital role in soil health and plant growth.

## Minerals

Minerals make up about 45% of soil composition and are derived from weathered rocks and minerals. They are categorized into three main particle sizes: sand, silt, and clay.

- **Sand:** Sand particles are the largest, measuring 0.05 to 2 mm in diameter. Sandy soils drain quickly and are often low in nutrients because they have a low capacity to hold water and nutrients.

- **Silt:** Silt particles are medium-sized, measuring 0.002 to 0.05 mm in diameter. Silty soils feel smooth and can retain more moisture and nutrients than sandy soils but still drain relatively well.

- **Clay:** Clay particles are the smallest, measuring less than 0.002 mm in diameter. Clay soils can hold a significant amount of water and nutrients but can become compacted and poorly drained if not managed properly.

The relative proportions of these particles determine the soil texture, which can be classified into several types, such as sandy loam, silty clay, or clay loam. A balanced mixture of sand, silt, and clay is known as loam, which is ideal for most plants.

## Organic Matter

Organic matter, which constitutes about 5% of soil composition, includes decomposed plant and animal residues, living organisms, and humus (the stable end product of decomposed organic material). Organic matter is crucial for soil health as it improves soil structure, water retention, nutrient supply, and microbial activity.

## Water

Water occupies about 25% of the soil's volume in a well-balanced soil. It is essential for plant growth as it dissolves nutrients that plants absorb through their roots. The availability of water in soil depends on its texture and structure.

# Air

Air fills the remaining 25% of the soil's volume. It is vital for the respiration of plant roots and soil organisms. Well-aerated soil allows for the exchange of gases, including oxygen, carbon dioxide, and nitrogen, which are necessary for plant and microbial life.

# Soil Structure

Soil structure refers to the arrangement of soil particles into aggregates or clumps. Good soil structure enhances water infiltration, root penetration, and air exchange. Soil structure can be influenced by factors such as organic matter content, soil organisms, and management practices.

## Types of Soil Structure

- **Granular:** Small, rounded aggregates, typically found in the surface layers of well-aerated, organically rich soils. Ideal for plant growth.

- **Blocky:** Irregularly shaped aggregates that fit together like building blocks. Common in subsoils with higher clay content.

- **Platy:** Thin, flat plates of soil that lie horizontally. Often a result of compaction and poor drainage.

- **Prismatic and Columnar:** Vertical columns or prisms of soil, usually found in subsoils with high clay content and poor drainage.

- **Single Grain:** Individual soil particles that do not form aggregates, typical of sandy soils.

## Testing and Improving Soil Health

Regular soil testing and improvement are essential for maintaining optimal soil health and productivity. Soil tests provide valuable information about soil composition, nutrient levels, pH, and other characteristics.

## Soil Testing

### Why Test Soil?

Soil testing helps identify nutrient deficiencies, imbalances, and other issues that may affect plant growth. It allows you to tailor soil amendments and management practices to meet the specific needs of your soil and crops.

### When to Test Soil?

Soil testing should be done before establishing a new garden or planting area and periodically thereafter, ideally every 2-3 years. Testing should be done in the fall or early spring when the soil is neither too wet nor too dry.

### How to Collect Soil Samples

1. **Gather Tools:** You will need a clean bucket, a trowel or soil probe, and a soil sample bag or container.

2. **Select Sampling Sites:** Choose several locations within the area to be tested. Avoid areas with unusual features, such as compost piles, animal pens, or recently fertilized spots.

3. **Collect Samples:** Remove any surface debris. Using the trowel

or soil probe, collect soil samples from a depth of 6-8 inches. Take multiple samples (10-15) from different spots and mix them thoroughly in the bucket to create a composite sample.

4. **Prepare Sample for Testing:** Allow the composite sample to air dry. Remove any large debris or stones. Place the dried sample in a soil sample bag or container and label it with relevant information (e.g., location, date).

5. **Submit Sample for Testing:** Send the sample to a reputable soil testing laboratory. Follow the lab's instructions for sample submission and specify any particular tests or analyses needed.

## Interpreting Soil Test Results

Soil test results typically include information about soil pH, nutrient levels (macronutrients and micronutrients), organic matter content, and recommendations for amendments.

- **pH:** Measures soil acidity or alkalinity. Most plants prefer a pH of 6.0-7.0. Adjust pH using lime (to raise pH) or sulfur (to lower pH) based on recommendations.

- **Macronutrients:** Essential nutrients required in larger quantities, including nitrogen (N), phosphorus (P), and potassium (K). Soil test results will indicate if these nutrients are deficient, sufficient, or excessive.

- **Micronutrients:** Essential nutrients required in smaller quantities, including iron (Fe), manganese (Mn), zinc (Zn), copper (Cu), boron (B), molybdenum (Mo), and chlorine (Cl).

- **Organic Matter:** Indicates the amount of decomposed organic material in the soil. Higher organic matter content generally improves soil health and fertility.

## Improving Soil Health

Improving soil health involves enhancing its physical, chemical, and biological properties. Here are several practices to improve soil health:

## Adding Organic Matter

Incorporating organic matter into the soil improves its structure, water retention, nutrient availability, and microbial activity. Sources of organic matter include compost, manure, cover crops, and mulch.

- **Compost:** Composting transforms organic waste into a nutrient-rich soil amendment. Apply 1-2 inches of compost to the soil surface or incorporate it into the topsoil.

- **Manure:** Well-rotted animal manure adds nutrients and organic matter. Avoid using fresh manure, as it can burn plants and introduce pathogens.

- **Cover Crops:** Growing cover crops, such as legumes or grasses, adds organic matter, improves soil structure, and enhances nutrient cycling. Incorporate cover crops into the soil before they go to seed.

- **Mulch:** Applying organic mulch (e.g., straw, leaves, wood chips) helps retain soil moisture, suppress weeds, and add organic matter as it decomposes.

## Balancing Soil pH

Maintaining the correct soil pH is crucial for nutrient availability and plant health. Use lime to raise soil pH and sulfur to lower it, following soil test recommendations.

- **Lime:** Apply ground limestone (calcium carbonate) or dolomitic lime (calcium magnesium carbonate) to raise soil pH. The amount needed depends on soil type and current pH levels.

- **Sulfur:** Apply elemental sulfur or sulfate-containing compounds (e.g., ammonium sulfate) to lower soil pH. The amount needed depends on soil type and current pH levels.

## Enhancing Nutrient Availability

Based on soil test results, apply appropriate fertilizers and amendments to address nutrient deficiencies or imbalances. Organic fertilizers, such as compost, manure, and bone meal, are recommended for sustainable soil management.

- **Nitrogen (N):** Add composted manure, green manure, or nitrogen-fixing cover crops (e.g., clover, alfalfa) to increase nitrogen levels.

- **Phosphorus (P):** Apply bone meal, rock phosphate, or compost to increase phosphorus levels.

- **Potassium (K):** Use compost, wood ash, or greensand to increase potassium levels.

- **Micronutrients:** Apply specific amendments based on deficiencies (e.g., iron sulfate for iron deficiency, zinc sulfate for zinc deficiency).

## Improving Soil Structure

Enhancing soil structure involves practices that promote the formation of stable aggregates, increase pore space, and reduce compaction.

- **Organic Matter:** Regularly add organic matter to improve soil structure and aggregate stability.

- **Cover Crops:** Grow cover crops to protect soil from erosion, improve structure, and increase organic matter.

- **Reduced Tillage:** Minimize soil disturbance by using no-till or reduced-till practices to maintain soil structure and reduce compaction.

- **Mulching:** Apply mulch to protect soil from erosion, retain moisture, and add organic matter.

## Promoting Soil Biology

Healthy soil is teeming with beneficial microorganisms, fungi, earthworms, and other organisms that contribute to nutrient cycling, disease suppression, and soil structure.

- **Compost:** Regularly apply compost to introduce and support beneficial soil organisms.

- **Cover Crops:** Grow diverse cover crops to provide habitat and food for soil organisms.

- **Avoid Chemical Overuse:** Minimize the use of synthetic fertilizers, pesticides, and herbicides that can harm beneficial soil organisms.

- **Crop Rotation:** Rotate crops to prevent the buildup of pests and diseases, promote diverse root structures, and support soil health.

# Organic Soil Amendments

Organic soil amendments are materials derived from natural sources that are added to the soil to improve its physical, chemical, and biological properties. They play a vital role in sustainable soil management and enhance soil health and fertility.

## Types of Organic Soil Amendments

### Compost

Compost is decomposed organic matter that enriches soil with nutrients and beneficial microorganisms. It improves soil structure, water retention, and fertility.

- **Ingredients:** Compost is made from a mix of green materials (e.g., grass clippings, vegetable scraps) and brown materials (e.g., leaves, straw, wood chips).

- **Benefits:** Enhances soil structure, increases nutrient availability, improves water retention, and promotes beneficial microbial activity.

- **Application:** Apply 1-2 inches of compost to the soil surface or incorporate it into the topsoil. Use it in garden beds, around trees and shrubs, and in potting mixes.

## Manure

Animal manure is a rich source of organic matter and nutrients. It must be well-composted or aged to avoid burning plants and introducing pathogens.

- **Types:** Common types include cow, horse, chicken, and sheep manure. Each type varies in nutrient content.

- **Benefits:** Adds organic matter, provides a balanced mix of nutrients, improves soil structure, and promotes microbial activity.

- **Application:** Apply 2-3 inches of well-composted manure to the soil surface or incorporate it into the topsoil. Avoid using fresh manure directly on crops.

## Cover Crops

Cover crops are plants grown primarily to improve soil health. They add organic matter, fix nitrogen, suppress weeds, and prevent erosion.

- **Types:** Legumes (e.g., clover, vetch, peas) fix nitrogen, while grasses (e.g., rye, oats, barley) add biomass and protect soil.

- **Benefits:** Increase organic matter, improve soil structure, enhance nutrient cycling, and provide habitat for beneficial organisms.

- **Application:** Sow cover crops during fallow periods or as part of crop rotation. Incorporate them into the soil before they go to seed.

## Mulch

Mulch is a layer of organic material applied to the soil surface. It conserves moisture, suppresses weeds, regulates soil temperature, and adds organic matter as it decomposes.

- **Types:** Common mulches include straw, leaves, grass clippings, wood chips, and bark.

- **Benefits:** Reduces evaporation, suppresses weeds, moderates soil temperature, and adds organic matter.

- **Application:** Apply a 2-4 inch layer of mulch around plants, trees, and shrubs. Replenish mulch as it decomposes.

## Green Manure

Green manure is a type of cover crop grown specifically to be incorporated into the soil while still green. It adds organic matter and nutrients.

- **Types:** Common green manures include clover, alfalfa, buckwheat, and mustard.

- **Benefits:** Adds organic matter, improves soil structure, increases nutrient availability, and suppresses weeds.

- **Application:** Sow green manure crops during fallow periods or between main crops. Incorporate them into the soil before they flower.

## Bone Meal

Bone meal is a finely ground powder made from animal bones. It is a slow-release source of phosphorus and calcium.

- **Benefits:** Provides a long-lasting source of phosphorus, essential for root development and flowering. Also supplies calcium, important for cell structure and strength.

- **Application:** Apply bone meal at planting time by mixing it into the soil at the root zone. Follow package recommendations for application rates.

## Blood Meal

Blood meal is a dry, powdered form of animal blood. It is a high-nitrogen organic fertilizer.

- **Benefits:** Provides a quick-release source of nitrogen, essential for vegetative growth. Also helps to repel certain pests.

- **Application:** Apply blood meal to the soil surface or mix it into the topsoil. Use caution, as it can burn plants if over-applied. Follow package recommendations for application rates.

## Fish Emulsion

Fish emulsion is a liquid fertilizer made from processed fish. It is rich in nitrogen and other nutrients.

- **Benefits:** Provides a readily available source of nitrogen, phosphorus, and potassium. Also contains trace minerals and beneficial microbes.

- **Application:** Dilute fish emulsion with water according to package instructions and apply as a foliar spray or soil drench. Use it as a quick boost for plants during the growing season.

## Seaweed Extract

Seaweed extract is a liquid or powdered fertilizer made from marine algae. It is a source of micronutrients, growth hormones, and beneficial compounds.

- **Benefits:** Provides trace minerals, enhances plant growth, improves stress tolerance, and promotes microbial activity.

- **Application:** Dilute seaweed extract with water according to package instructions and apply as a foliar spray or soil drench. Use it to supplement regular fertilization and support plant health.

## Benefits of Organic Soil Amendments

## Enhanced Soil Structure

Organic amendments improve soil structure by increasing the formation of aggregates, enhancing porosity, and reducing compaction. Improved soil structure promotes root growth, water infiltration, and air exchange.

## Increased Nutrient Availability

Organic amendments provide essential nutrients, both macronutrients (nitrogen, phosphorus, potassium) and micronutrients (iron, manganese, zinc), in a slow-release form. They also enhance nutrient cycling and availability through microbial activity.

## Improved Water Retention

Adding organic matter increases the soil's ability to retain moisture, reducing the need for frequent irrigation. This is particularly beneficial in sandy soils that drain quickly and in drought-prone areas.

## Enhanced Microbial Activity

Organic amendments support a diverse and active soil microbial community. Beneficial microbes decompose organic matter, release nutrients, suppress pathogens, and improve soil structure.

## Long-Term Soil Fertility

Organic amendments build long-term soil fertility by increasing organic matter content, improving nutrient reserves, and enhancing soil health. They contribute to sustainable soil management and reduce dependence on synthetic fertilizers.

## Application Tips for Organic Soil Amendments

1. **Soil Testing:** Conduct a soil test before applying amendments to identify nutrient deficiencies and pH imbalances. Use test results to guide amendment selection and application rates.

2. **Timing:** Apply organic amendments at the right time to maximize their benefits. For example, apply compost and manure in the fall or early spring to allow for decomposition and nutrient release before planting.

3. **Incorporation:** Incorporate amendments into the topsoil to enhance their effectiveness. Use a garden fork, tiller, or shovel to mix them thoroughly into the soil.

4. **Cover Crops:** Grow cover crops during fallow periods or as part of crop rotation to add organic matter and improve soil health. Incorporate cover crops into the soil before they go to seed.

5. **Mulching:** Apply organic mulch to the soil surface to conserve moisture, suppress weeds, and add organic matter as it decomposes. Replenish mulch as needed.

6. **Rotation:** Rotate the use of different amendments to prevent nutrient imbalances and support diverse soil biology. For example, alternate between compost, manure, and green manure.

7. **Avoid Over-Application:** Follow recommended application rates to avoid nutrient imbalances and potential negative effects on soil and plant health. Over-application can lead to nutrient leaching, pollution, and plant damage.

Understanding soil composition and structure, testing and improving soil health, and utilizing organic soil amendments are fundamental to successful homesteading. By focusing on soil health, you create a fertile and productive foundation for your homestead. Regular soil testing, the addition of organic matter, and the use of sustainable soil management practices will ensure that your soil remains healthy and productive for years to come. By investing in your soil, you are investing in the long-term success and sustainability of your homestead.

Composting is a natural process of recycling organic materials, such as leaves and vegetable scraps, into a rich soil amendment known as compost. This process involves the breakdown of organic matter by microorganisms under controlled conditions. Composting not only reduces waste but also improves soil health by adding valuable nutrients and enhancing soil structure. In this section, we will explore the science of composting, how to build a compost pile, and the practice of vermiculture, or worm composting.

# Composting

## The Science of Composting

### Biological Processes

Composting is driven by a variety of microorganisms, primarily bacteria, fungi, and actinomycetes, which decompose organic matter into simpler compounds. These microorganisms require specific conditions to thrive and efficiently break down organic materials.

### Bacteria

Bacteria are the most abundant microorganisms in a compost pile and are primarily responsible for the decomposition process. There are different types of bacteria involved in composting, categorized by their temperature preferences:

- **Psychrophilic Bacteria:** These bacteria thrive at low temperatures (32-55°F). They initiate the composting process in the early stages when the pile is still cool.

- **Mesophilic Bacteria:** These bacteria operate at moderate temperatures (70-100°F). They take over from psychrophilic bacteria as the pile heats up and continue the decomposition process.

- **Thermophilic Bacteria:** These heat-loving bacteria thrive at high temperatures (104-160°F). They become active as the pile heats further and are responsible for the rapid breakdown of organic matter. Thermophilic bacteria also help to kill pathogens and weed seeds due to the high temperatures they generate.

## Fungi

Fungi, including molds and yeasts, play a secondary role in composting. They break down complex organic compounds like lignin and cellulose, which are more resistant to decomposition by bacteria. Fungi are particularly active in the later stages of composting when the pile begins to cool.

## Actinomycetes

Actinomycetes are a group of bacteria that resemble fungi in their filamentous structure. They are essential in breaking down tough organic materials like woody stems and bark. Actinomycetes thrive in the later stages of composting and are responsible for the earthy smell of finished compost.

## Physical and Chemical Processes

In addition to biological activity, several physical and chemical processes are involved in composting. These processes help to transform organic matter into stable, nutrient-rich compost.

## Aeration

Oxygen is crucial for the aerobic decomposition process. Aeration ensures that composting microorganisms have enough oxygen to efficiently break down organic materials. Turning the compost pile periodically helps to maintain adequate oxygen levels and prevents the pile from becoming anaerobic, which can lead to foul odors and slower decomposition.

## Moisture

Moisture is another critical factor in composting. Microorganisms require water to survive and function effectively. The ideal moisture content for a compost pile is around 50-60%. If the pile is too dry, decomposition will slow down; if it is too wet, it can become anaerobic and produce unpleasant smells.

## Temperature

Temperature is an important indicator of microbial activity in a compost pile. As microorganisms break down organic matter, they generate heat. A well-managed compost pile can reach temperatures of 130-160°F during the thermophilic phase. Maintaining high temperatures for several days helps to kill pathogens and weed seeds. As decomposition slows, the pile gradually cools to ambient temperature.

# Carbon-to-Nitrogen Ratio (C

## Ratio)

The carbon-to-nitrogen ratio (C

ratio) is a key factor in composting. Carbon provides energy for microorganisms, while nitrogen is essential for their growth and reproduction. The optimal C

ratio for composting is around 30:1. Materials high in carbon, known as "browns," include leaves, straw, and wood chips. Materials high in nitrogen, known as "greens," include vegetable scraps, grass clippings, and manure. Balancing these materials ensures efficient decomposition and prevents issues such as odor or slow breakdown.

## Stages of Composting

Composting progresses through several stages, each characterized by different temperatures and microbial activities:

1. **Mesophilic Stage:** This initial stage lasts a few days, with temperatures rising to 70-100°F. Mesophilic bacteria break down easily degradable materials.

2. **Thermophilic Stage:** Lasting several weeks, this stage sees temperatures rise to 104-160°F. Thermophilic bacteria dominate, rapidly decomposing organic matter and sanitizing the compost by killing pathogens and weed seeds.

3. **Cooling Stage:** As the easily degradable materials are exhausted, the

temperature begins to drop. Mesophilic bacteria and fungi become more active again, breaking down more resistant materials.

4. **Maturation Stage:** This final stage can last several months. The compost continues to stabilize and mature, with the formation of humus. Actinomycetes and fungi play significant roles during this stage.

## Building a Compost Pile

### Selecting a Site

Choosing the right location for your compost pile is crucial for its success. Consider the following factors:

- **Accessibility:** Choose a site that is easily accessible for adding materials and turning the pile.

- **Drainage:** Ensure the site has good drainage to prevent waterlogging.

- **Sunlight:** A balance of sun and shade helps maintain optimal temperatures and moisture levels.

- **Proximity to Garden:** Place the compost pile near your garden for convenience in adding finished compost to your soil.

### Choosing a Composting System

There are various composting systems to choose from, each with its advantages and disadvantages:

## Open Pile

An open pile is the simplest composting system, consisting of a heap of organic materials. It requires regular turning and monitoring but is cost-effective and easy to set up.

- **Advantages:** Low cost, simple setup, suitable for large quantities of materials.

- **Disadvantages:** Requires frequent turning, may attract pests, and can be visually unappealing.

## Compost Bin

Compost bins are enclosed structures that contain the composting materials. They come in various designs, including stationary bins, tumblers, and DIY options.

- **Advantages:** Contain materials neatly, reduce pest problems, and may retain heat and moisture better than open piles.

- **Disadvantages:** Can be more expensive, may require assembly, and limit the amount of material that can be composted at one time.

## Tumblers

Tumblers are rotating compost bins that simplify the turning process. They are particularly useful for smaller spaces and quicker composting.

- **Advantages:** Easy to turn, fast composting, pest-resistant, and contained.

- **Disadvantages:** Limited capacity, can be expensive, and may require physical effort to rotate.

## Gathering Materials

Successful composting requires a mix of "browns" (carbon-rich materials) and "greens" (nitrogen-rich materials).

## Browns (Carbon-Rich Materials)

- Leaves

- Straw

- Wood chips

- Sawdust

- Shredded paper and cardboard

## Greens (Nitrogen-Rich Materials)

- Vegetable scraps

- Fruit peels

- Grass clippings

- Coffee grounds

- Manure (from herbivores like cows, horses, and rabbits)

## Building the Pile

Building a compost pile involves layering materials and maintaining the right balance of carbon, nitrogen, moisture, and oxygen.

1. **Layering:** Start with a layer of coarse materials like straw or wood chips to improve aeration and drainage. Alternate layers of browns and greens, aiming for a Cratio of around 30:1. Add water to each layer to maintain moisture.

2. **Size:** The ideal size for a compost pile is at least 3 feet wide, 3 feet deep, and 3 feet high. This size ensures adequate heat retention while allowing for sufficient aeration.

3. **Turning:** Turn the pile regularly to introduce oxygen and speed up decomposition. Aim to turn the pile every 1-2 weeks, or when the temperature begins to drop.

## Monitoring and Maintaining the Pile

Regular monitoring and maintenance are key to successful composting. Check the pile for temperature, moisture, and odor.

- **Temperature:** Use a compost thermometer to monitor the pile's temperature. Aim for 130-160°F during the thermophilic stage. If the temperature drops, turn the pile to reintroduce oxygen and mix in fresh materials if needed.

- **Moisture:** The pile should be as moist as a wrung-out sponge. If it is too dry, add water or more green materials. If it is too wet, add more brown materials and turn the pile to improve aeration.

- **Odor:** A well-maintained compost pile should have an earthy smell. If it smells rotten or ammonia-like, it may be too wet or have too much nitrogen. Adjust the balance of materials and turn the pile to resolve the issue.

## Harvesting Finished Compost

Finished compost is dark, crumbly, and has an earthy smell. Depending on the composting method and materials used, it can take anywhere from a few months to a year to produce finished compost.

- **Screening:** Use a compost screen or sieve to separate finished compost from larger, undecomposed materials. Return the larger materials to the pile for further decomposition.

- **Storage:** Store finished compost in a covered bin or pile until you are ready to use it. Keep it moist to maintain its microbial activity.

- **Application:** Apply compost to garden beds, around trees and shrubs, or mix it into potting soil to improve soil fertility and structure.

# Vermiculture: Worm Composting

Vermiculture, or worm composting, is a method of composting that uses worms to break down organic matter into nutrient-rich castings. This process is particularly suitable for kitchen scraps and can be done indoors or outdoors.

## Benefits of Vermiculture

- **Efficiency:** Worms can process organic matter quickly, producing compost in as little as 2-3 months.

- **Space:** Vermiculture requires less space than traditional composting, making it ideal for small gardens or indoor use.

- **Odor:** Properly managed worm bins have little to no odor.

- **Nutrient-Rich:** Worm castings are highly nutrient-rich and beneficial for plant growth.

## Setting Up a Worm Bin

### Choosing a Container

A worm bin can be made from a variety of materials, including plastic, wood, or metal. The container should be opaque to keep worms in the dark and have a lid to retain moisture and exclude pests.

- **Size:** For a small household, a bin that is 2-3 feet wide, 1-2 feet deep, and 1-2 feet high is sufficient. Larger bins can be used for greater quantities of waste.

- **Drainage:** Ensure the bin has drainage holes to prevent waterlogging. Place a tray or saucer underneath to catch any excess liquid, known as "worm tea," which can be used as a liquid fertilizer.

- **Ventilation:** Drill ventilation holes near the top of the bin to allow for air circulation.

### Bedding Materials

Bedding provides a habitat for worms and helps maintain moisture and aeration. Suitable bedding materials include:

- Shredded newspaper or cardboard

- Straw

- Peat moss

- Coconut coir

- A small amount of soil or compost to introduce beneficial microorganisms

Moisten the bedding materials until they are as damp as a wrung-out sponge, then fill the bin about two-thirds full.

## Choosing and Adding Worms

The best worms for composting are red wigglers (Eisenia fetida) or redworms (Lumbricus rubellus). These species thrive in composting conditions and process organic matter efficiently.

- **Quantity:** For a small bin, start with about 1 pound of worms (approximately 1,000 worms). This amount can process about half a pound of food waste per day.

- **Introduction:** Gently spread the worms on top of the bedding. They will quickly burrow down into the bedding to avoid light.

## Feeding the Worms

Worms consume a variety of organic materials, but some items are better suited for vermiculture than others.

## Suitable Food Scraps

- Fruit and vegetable scraps (avoid citrus and acidic foods)

- Coffee grounds and filters

- Tea bags (remove staples)

- Crushed eggshells

- Small amounts of bread and grains

## Avoid

- Meat, dairy, and oily foods

- Citrus and acidic foods

- Spicy foods

- Pet waste

- Non-biodegradable materials

Feed the worms by burying food scraps in the bedding. Rotate the feeding spots to ensure even decomposition and prevent odors.

## Maintaining the Worm Bin

Regular maintenance is essential for a healthy worm bin.

- **Moisture:** Check the moisture level of the bedding regularly. Add water if it becomes too dry, and add dry bedding if it becomes too wet.

- **Aeration:** Gently fluff the bedding occasionally to improve aeration and prevent compaction.

- **Temperature:** Keep the worm bin in a location where the temperature remains between 55-77°F. Avoid placing it in direct sunlight or near heat sources.

## Harvesting Worm Castings

Worm castings are ready to harvest when the bedding is dark and crumbly, and food scraps are no longer recognizable. There are several methods for harvesting castings:

- **Side-to-Side Method:** Move the contents of the bin to one side and add fresh bedding and food scraps to the empty side. The worms will migrate to the new bedding, allowing you to harvest the castings from the original side.

- **Light Method:** Spread the contents of the bin in small mounds on a flat surface under a bright light. Worms will burrow away from the light, allowing you to remove the top layer of castings. Repeat until mostly worms remain.

- **Screening:** Use a screen or sieve to separate castings from worms and larger materials. Return the worms and undecomposed materials to the bin.

## Using Worm Castings

Worm castings are a potent organic fertilizer that can be used in various ways:

- **Soil Amendment:** Mix castings into garden beds or potting soil to improve fertility and structure.

- **Top Dressing:** Sprinkle castings around the base of plants as a nutrient-rich mulch.

- **Compost Tea:** Steep castings in water to create a liquid fertilizer known as compost tea. Use it to water plants or as a foliar spray.

Composting and vermiculture are effective and sustainable ways to recycle organic waste and improve soil health. Understanding the science of composting, building a compost pile, and practicing vermiculture can help you create nutrient-rich compost for your homestead. By following the principles and techniques outlined in this section, you can reduce waste, enhance soil fertility, and contribute to a more sustainable and productive homestead.

# Chapter 3

# Composting

## Biological Processes

Composting is driven by a variety of microorganisms, primarily bacteria, fungi, and actinomycetes, which decompose organic matter into simpler compounds. These microorganisms require specific conditions to thrive and efficiently break down organic materials.

## Bacteria

Bacteria are the most abundant microorganisms in a compost pile and are primarily responsible for the decomposition process. There are different types of bacteria involved in composting, categorized by their temperature preferences:

- **Psychrophilic Bacteria:** These bacteria thrive at low temperatures (32-55°F). They initiate the composting process in the early stages when the pile is still cool.

- **Mesophilic Bacteria:** These bacteria operate at moderate temperatures (70-100°F). They take over from psychrophilic bacteria as the pile heats up and continue the decomposition process.

- **Thermophilic Bacteria:** These heat-loving bacteria thrive at high temperatures (104-160°F). They become active as the pile heats further and are responsible for the rapid breakdown of organic matter. Thermophilic bacteria also help to kill pathogens and weed seeds due to the high temperatures they generate.

## Fungi

Fungi, including molds and yeasts, play a secondary role in composting. They break down complex organic compounds like lignin and cellulose, which are more resistant to decomposition by bacteria. Fungi are particularly active in the later stages of composting when the pile begins to cool.

## Actinomycetes

Actinomycetes are a group of bacteria that resemble fungi in their filamentous structure. They are essential in breaking down tough organic materials like woody stems and bark. Actinomycetes thrive in the later stages of composting and are responsible for the earthy smell of finished compost.

## Physical and Chemical Processes

In addition to biological activity, several physical and chemical processes are involved in composting. These processes help to transform organic matter into stable, nutrient-rich compost.

## Aeration

Oxygen is crucial for the aerobic decomposition process. Aeration ensures that composting microorganisms have enough oxygen to efficiently break down organic materials. Turning the compost pile periodically helps to maintain adequate oxygen levels and prevents the pile from becoming anaerobic, which can lead to foul odors and slower decomposition.

## Moisture

Moisture is another critical factor in composting. Microorganisms require water to survive and function effectively. The ideal moisture content for a compost pile is around 50-60%. If the pile is too dry, decomposition will slow down; if it is too wet, it can become anaerobic and produce unpleasant smells.

## Temperature

Temperature is an important indicator of microbial activity in a compost pile. As microorganisms break down organic matter, they generate heat. A well-managed compost pile can reach temperatures of 130-160°F during the thermophilic phase. Maintaining high temperatures for several days helps to kill pathogens and weed seeds. As decomposition slows, the pile gradually cools to ambient temperature.

## Carbon-to-Nitrogen Ratio (C

## Ratio)

The carbon-to-nitrogen ratio (C

ratio) is a key factor in composting. Carbon provides energy for microorganisms, while nitrogen is essential for their growth and reproduction. The optimal C

ratio for composting is around 30:1. Materials high in carbon, known as "browns," include leaves, straw, and wood chips. Materials high in nitrogen, known as "greens," include vegetable scraps, grass clippings, and manure. Balancing these materials ensures efficient decomposition and prevents issues such as odor or slow breakdown.

## Stages of Composting

Composting progresses through several stages, each characterized by different temperatures and microbial activities:

1. **Mesophilic Stage:** This initial stage lasts a few days, with temperatures rising to 70-100°F. Mesophilic bacteria break down easily degradable materials.

2. **Thermophilic Stage:** Lasting several weeks, this stage sees temperatures rise to 104-160°F. Thermophilic bacteria dominate, rapidly decomposing organic matter and sanitizing the compost by killing pathogens and weed seeds.

3. **Cooling Stage:** As the easily degradable materials are exhausted, the temperature begins to drop. Mesophilic bacteria and fungi become more active again, breaking down more resistant materials.

4. **Maturation Stage:** This final stage can last several months. The compost continues to stabilize and mature, with the formation of humus. Actinomycetes and fungi play significant roles during this

stage.

# Building a Compost Pile

## Selecting a Site

Choosing the right location for your compost pile is crucial for its success. Consider the following factors:

- **Accessibility:** Choose a site that is easily accessible for adding materials and turning the pile.

- **Drainage:** Ensure the site has good drainage to prevent waterlogging.

- **Sunlight:** A balance of sun and shade helps maintain optimal temperatures and moisture levels.

- **Proximity to Garden:** Place the compost pile near your garden for convenience in adding finished compost to your soil.

## Choosing a Composting System

There are various composting systems to choose from, each with its advantages and disadvantages:

## Open Pile

An open pile is the simplest composting system, consisting of a heap of organic materials. It requires regular turning and monitoring but is cost-effective and easy to set up.

- **Advantages:** Low cost, simple setup, suitable for large quantities of materials.

- **Disadvantages:** Requires frequent turning, may attract pests, and can be visually unappealing.

## Compost Bin

Compost bins are enclosed structures that contain the composting materials. They come in various designs, including stationary bins, tumblers, and DIY options.

- **Advantages:** Contain materials neatly, reduce pest problems, and may retain heat and moisture better than open piles.

- **Disadvantages:** Can be more expensive, may require assembly, and limit the amount of material that can be composted at one time.

## Tumblers

Tumblers are rotating compost bins that simplify the turning process. They are particularly useful for smaller spaces and quicker composting.

- **Advantages:** Easy to turn, fast composting, pest-resistant, and contained.

- **Disadvantages:** Limited capacity, can be expensive, and may require physical effort to rotate.

## Gathering Materials

Successful composting requires a mix of "browns" (carbon-rich materials) and "greens" (nitrogen-rich materials).

## Browns (Carbon-Rich Materials)

- Leaves

- Straw

- Wood chips

- Sawdust

- Shredded paper and cardboard

## Greens (Nitrogen-Rich Materials)

- Vegetable scraps

- Fruit peels

- Grass clippings

- Coffee grounds

- Manure (from herbivores like cows, horses, and rabbits)

## Building the Pile

Building a compost pile involves layering materials and maintaining the right balance of carbon, nitrogen, moisture, and oxygen.

1. **Layering:** Start with a layer of coarse materials like straw or wood

chips to improve aeration and drainage. Alternate layers of browns and greens, aiming for a Cratio of around 30:1. Add water to each layer to maintain moisture.

2. **Size:** The ideal size for a compost pile is at least 3 feet wide, 3 feet deep, and 3 feet high. This size ensures adequate heat retention while allowing for sufficient aeration.

3. **Turning:** Turn the pile regularly to introduce oxygen and speed up decomposition. Aim to turn the pile every 1-2 weeks, or when the temperature begins to drop.

## Monitoring and Maintaining the Pile

Regular monitoring and maintenance are key to successful composting. Check the pile for temperature, moisture, and odor.

- **Temperature:** Use a compost thermometer to monitor the pile's temperature. Aim for 130-160°F during the thermophilic stage. If the temperature drops, turn the pile to reintroduce oxygen and mix in fresh materials if needed.

- **Moisture:** The pile should be as moist as a wrung-out sponge. If it is too dry, add water or more green materials. If it is too wet, add more brown materials and turn the pile to improve aeration.

- **Odor:** A well-maintained compost pile should have an earthy smell. If it smells rotten or ammonia-like, it may be too wet or have too much nitrogen. Adjust the balance of materials and turn the pile to resolve the issue.

# Harvesting Finished Compost

Finished compost is dark, crumbly, and has an earthy smell. Depending on the composting method and materials used, it can take anywhere from a few months to a year to produce finished compost.

- **Screening:** Use a compost screen or sieve to separate finished compost from larger, undecomposed materials. Return the larger materials to the pile for further decomposition.

- **Storage:** Store finished compost in a covered bin or pile until you are ready to use it. Keep it moist to maintain its microbial activity.

- **Application:** Apply compost to garden beds, around trees and shrubs, or mix it into potting soil to improve soil fertility and structure.

# Vermiculture: Worm Composting

Vermiculture, or worm composting, is a method of composting that uses worms to break down organic matter into nutrient-rich castings. This process is particularly suitable for kitchen scraps and can be done indoors or outdoors.

## Benefits of Vermiculture

- **Efficiency:** Worms can process organic matter quickly, producing compost in as little as 2-3 months.

- **Space:** Vermiculture requires less space than traditional composting, making it ideal for small gardens or indoor use.

- **Odor:** Properly managed worm bins have little to no odor.

- **Nutrient-Rich:** Worm castings are highly nutrient-rich and beneficial for plant growth.

## Setting Up a Worm Bin

## Choosing a Container

A worm bin can be made from a variety of materials, including plastic, wood, or metal. The container should be opaque to keep worms in the dark and have a lid to retain moisture and exclude pests.

- **Size:** For a small household, a bin that is 2-3 feet wide, 1-2 feet deep, and 1-2 feet high is sufficient. Larger bins can be used for greater quantities of waste.

- **Drainage:** Ensure the bin has drainage holes to prevent waterlogging. Place a tray or saucer underneath to catch any excess liquid, known as "worm tea," which can be used as a liquid fertilizer.

- **Ventilation:** Drill ventilation holes near the top of the bin to allow for air circulation.

## Bedding Materials

Bedding provides a habitat for worms and helps maintain moisture and aeration. Suitable bedding materials include:

- Shredded newspaper or cardboard

- Straw

- Peat moss

- Coconut coir

- A small amount of soil or compost to introduce beneficial microorganisms

Moisten the bedding materials until they are as damp as a wrung-out sponge, then fill the bin about two-thirds full.

## Choosing and Adding Worms

The best worms for composting are red wigglers (Eisenia fetida) or redworms (Lumbricus rubellus). These species thrive in composting conditions and process organic matter efficiently.

- **Quantity:** For a small bin, start with about 1 pound of worms (approximately 1,000 worms). This amount can process about half a pound of food waste per day.

- **Introduction:** Gently spread the worms on top of the bedding. They will quickly burrow down into the bedding to avoid light.

## Feeding the Worms

Worms consume a variety of organic materials, but some items are better suited for vermiculture than others.

## Suitable Food Scraps

- Fruit and vegetable scraps (avoid citrus and acidic foods)

- Coffee grounds and filters

- Tea bags (remove staples)

- Crushed eggshells

- Small amounts of bread and grains

## Avoid

- Meat, dairy, and oily foods

- Citrus and acidic foods

- Spicy foods

- Pet waste

- Non-biodegradable materials

Feed the worms by burying food scraps in the bedding. Rotate the feeding spots to ensure even decomposition and prevent odors.

## Maintaining the Worm Bin

Regular maintenance is essential for a healthy worm bin.

- **Moisture:** Check the moisture level of the bedding regularly. Add water if it becomes too dry, and add dry bedding if it becomes too wet.

- **Aeration:** Gently fluff the bedding occasionally to improve aeration and prevent compaction.

- **Temperature:** Keep the worm bin in a location where the temperature remains between 55-77°F. Avoid placing it in direct sunlight or near heat sources.

## Harvesting Worm Castings

Worm castings are ready to harvest when the bedding is dark and crumbly, and food scraps are no longer recognizable. There are several methods for harvesting castings:

- **Side-to-Side Method:** Move the contents of the bin to one side and add fresh bedding and food scraps to the empty side. The worms will migrate to the new bedding, allowing you to harvest the castings from the original side.

- **Light Method:** Spread the contents of the bin in small mounds on a flat surface under a bright light. Worms will burrow away from the light, allowing you to remove the top layer of castings. Repeat until mostly worms remain.

- **Screening:** Use a screen or sieve to separate castings from worms and larger materials. Return the worms and undecomposed materials to the bin.

## Using Worm Castings

Worm castings are a potent organic fertilizer that can be used in various ways:

- **Soil Amendment:** Mix castings into garden beds or potting soil to improve fertility and structure.

- **Top Dressing:** Sprinkle castings around the base of plants as a nutrient-rich mulch.

- **Compost Tea:** Steep castings in water to create a liquid fertilizer known as compost tea. Use it to water plants or as a foliar spray.

Composting and vermiculture are effective and sustainable ways to recycle organic waste and improve soil health. Understanding the science of composting, building a compost pile, and practicing vermiculture can help you create nutrient-rich compost for your homestead. By following the principles and techniques outlined in this section, you can reduce waste, enhance soil fertility, and contribute to a more sustainable and productive homestead.

# Chapter 4

# Raised Beds and No-Till Gardening

Raised beds and no-till gardening are two highly effective methods for growing vegetables and other plants. Both techniques offer numerous benefits, including improved soil health, better water retention, and reduced weed pressure. This section will explore the benefits of raised beds, how to build and maintain them, and the techniques and advantages of no-till gardening.

## Benefits of Raised Beds

Raised beds are essentially garden plots that are elevated above the surrounding soil level. They can be constructed using various materials such as wood, stone, or metal and filled with a high-quality soil mix. The advantages of raised beds are numerous and can significantly enhance your gardening experience.

## Improved Soil Quality and Structure

One of the primary benefits of raised beds is the ability to control the soil quality. Since the beds are filled with soil that you choose, you can create the perfect growing medium for your plants. This is particularly beneficial if the native soil in your area is poor or contaminated.

- **Looser Soil:** Raised beds prevent soil compaction, providing a loose and friable soil structure that promotes root growth and improves aeration.

- **Better Drainage:** The elevation allows excess water to drain more effectively, preventing waterlogging and root rot.

## Extended Growing Season

Raised beds warm up more quickly in the spring than in-ground beds, allowing you to start planting earlier. The soil also stays warmer longer into the fall, extending your growing season.

- **Early Planting:** You can plant cool-season crops earlier in the spring.

- **Frost Protection:** Raised beds can be easily covered with row covers or plastic to protect against early and late frosts.

## Easier Maintenance

Raised beds make gardening tasks like planting, weeding, and harvesting more convenient, reducing the need for bending and kneeling.

- **Accessibility:** The elevated height is easier on the back and knees, making it ideal for older gardeners or those with physical limitations.

- **Defined Edges:** Raised beds create defined edges, keeping pathways clear and reducing weed encroachment.

## Pest and Weed Control

The structure of raised beds can help manage pests and weeds more effectively.

- **Fewer Weeds:** By starting with a clean soil mix and using mulch, you can significantly reduce weed growth.

- **Pest Barriers:** Physical barriers such as row covers, bird netting, or fencing can be easily installed to protect plants from pests.

## Space Efficiency

Raised beds can be constructed to fit into various spaces, making them suitable for small gardens, urban environments, and intensive planting techniques.

- **Intensive Planting:** Plants can be spaced more closely together, maximizing yield per square foot.

- **Customizable Sizes:** Raised beds can be built to fit any available space, from small patios to large garden plots.

## Building and Maintaining Raised Beds

Building and maintaining raised beds involves several steps, including selecting materials, choosing the right location, preparing the soil, and implementing proper maintenance practices. Here's a detailed guide to help you get started.

## Selecting Materials

Raised beds can be constructed from a variety of materials, each with its pros and cons. Your choice will depend on factors such as budget, durability, and aesthetics.

## Wood

Wood is a popular choice for raised beds due to its availability, ease of use, and natural look.

- **Cedar and Redwood:** These are rot-resistant and long-lasting, making them excellent choices for raised beds.

- **Pine or Fir:** These are more affordable but less durable than cedar or redwood. Treating them with non-toxic sealants can extend their lifespan.

- **Pressure-Treated Wood:** Modern pressure-treated wood is generally safe for use in vegetable gardens, but if you are concerned about chemicals, line the inside of the bed with landscape fabric.

## Stone and Brick

Stone and brick offer durability and a classic look but can be more expensive and labor-intensive to install.

- **Natural Stone:** Provides a rustic, long-lasting structure but can be costly and requires more skill to construct.

- **Concrete Blocks:** Affordable and easy to work with, but they can look industrial and may need a lining to prevent soil contamination.

## Metal

Metal raised beds are durable and have a modern look. They are resistant to rot and pests.

- **Galvanized Steel:** Long-lasting and rust-resistant, though it can heat up quickly in hot climates.

- **Corrugated Metal:** Provides a unique, modern aesthetic and is durable, though it may require a wooden or metal frame for support.

## Choosing the Right Location

Selecting the right location for your raised beds is crucial for plant health and productivity.

- **Sunlight:** Choose a location that receives at least 6-8 hours of direct sunlight per day. Most vegetables require full sun to thrive.

- **Accessibility:** Ensure the beds are easily accessible from all sides for planting, weeding, and harvesting. Leave enough space between beds for comfortable movement.

- **Water Source:** Position the beds near a water source for convenient irrigation.

## Preparing the Soil

The quality of soil in raised beds is paramount to their success. Using a well-balanced soil mix will ensure your plants get the nutrients they need.

## Soil Mix

A good soil mix for raised beds typically includes a combination of garden soil, compost, and soilless growing mediums.

- **Garden Soil:** Provides structure and minerals but should not make up more than 50% of the mix to avoid compaction.

- **Compost:** Adds organic matter and nutrients. Aim for about 25-50% of the mix.

- **Soilless Growing Mediums:** Materials like peat moss, coconut coir, or perlite improve aeration and drainage. Add these to make up about 25-50% of the mix.

## Soil Amendments

Amending the soil can improve its fertility and structure. Consider adding:

- **Organic Matter:** Regularly add compost or well-rotted manure to maintain soil fertility.

- **Mulch:** Use organic mulch such as straw, leaves, or wood chips to retain moisture, suppress weeds, and add nutrients as it decomposes.

## Building the Raised Bed

Building a raised bed involves a few straightforward steps. Here's a basic guide:

1. **Measure and Mark the Site:** Determine the dimensions of your bed and mark the location using stakes and string.

2. **Prepare the Ground:** Remove any grass or weeds from the site. If the soil is compacted, loosen it with a garden fork to improve drainage.

3. **Assemble the Frame:** Cut the materials to size and assemble the frame using screws or nails. Ensure the corners are square.

4. **Level the Frame:** Place the frame in the prepared location and ensure it is level. Adjust as necessary.

5. **Fill the Bed:** Fill the frame with the prepared soil mix. Water the soil lightly to help it settle, then add more soil if needed.

## Maintaining Raised Beds

Proper maintenance will ensure the longevity and productivity of your raised beds.

## Watering

Raised beds can dry out more quickly than in-ground gardens, so consistent watering is crucial.

- **Drip Irrigation:** Install a drip irrigation system to provide consistent, efficient watering.

- **Mulching:** Mulch the soil surface to retain moisture and reduce evaporation.

## Fertilizing

Regularly feeding your plants will promote healthy growth and productivity.

- **Organic Fertilizers:** Use compost, worm castings, or organic granular fertilizers to add nutrients to the soil.

- **Soil Testing:** Periodically test the soil to monitor nutrient levels and pH. Adjust fertilization as needed based on test results.

## Weeding

Weeds can compete with your plants for nutrients and water.

- **Mulch:** Mulch helps suppress weeds by blocking light to weed seeds.

- **Hand Weeding:** Regularly hand weed to remove any emerging weeds before they become established.

## Pest and Disease Management

Monitor your raised beds for pests and diseases.

- **Companion Planting:** Planting certain crops together can help deter pests and promote healthy growth.

- **Physical Barriers:** Use row covers, netting, or fencing to protect plants from pests.

- **Organic Controls:** Use organic pesticides or homemade solutions to manage pests and diseases.

# No-Till Gardening Techniques

No-till gardening, also known as no-dig gardening, is a method that avoids disturbing the soil through tillage. Instead, organic matter is added to the soil surface, and natural processes are allowed to improve soil structure and fertility over time.

## Benefits of No-Till Gardening

No-till gardening offers numerous advantages, particularly for soil health and sustainability.

### Improved Soil Structure

Avoiding tillage helps maintain soil structure and promotes the development of healthy soil aggregates.

- **Less Compaction:** Tillage can compact soil layers, reducing pore space and impeding root growth. No-till methods preserve soil porosity.

- **Enhanced Aeration:** Undisturbed soil has better air and water infiltration, supporting root and microbial health.

### Increased Organic Matter

Adding organic matter to the soil surface continuously improves soil fertility and structure.

- **Mulching:** Applying organic mulch helps retain moisture, suppress weeds, and add nutrients as it decomposes.

- **Cover Crops:** Growing cover crops during off-seasons adds organic matter and protects soil from erosion.

## Enhanced Soil Biology

No-till gardening supports a diverse and active soil ecosystem.

- **Microbial Activity:** Undisturbed soil promotes the growth of beneficial microorganisms that help decompose organic matter and cycle nutrients.

- **Earthworms and Other Soil Fauna:** No-till practices create a habitat for earthworms and other beneficial organisms that improve soil structure and fertility.

## Reduced Erosion

By maintaining soil cover and structure, no-till gardening minimizes soil erosion.

- **Soil Cover:** Mulch and cover crops protect the soil surface from wind and water erosion.

- **Stable Aggregates:** Healthy soil structure reduces the risk of soil particles being washed or blown away.

# Techniques for No-Till Gardening

Implementing no-till gardening involves several key practices and techniques to manage soil health and productivity.

## Mulching

Mulching is a cornerstone of no-till gardening, providing numerous benefits for soil and plant health.

- **Types of Mulch:** Organic mulches such as straw, leaves, grass clippings, wood chips, and compost are ideal. Avoid using materials that may introduce weed seeds or disease.

- **Application:** Apply a thick layer (2-4 inches) of mulch around plants and over bare soil. Replenish mulch as it decomposes.

- **Benefits:** Mulch retains moisture, suppresses weeds, regulates soil temperature, and adds organic matter to the soil.

## Cover Cropping

Cover cropping involves growing specific plants during off-seasons to protect and improve the soil.

- **Types of Cover Crops:** Common cover crops include legumes (e.g., clover, vetch), grasses (e.g., rye, oats), and brassicas (e.g., mustard).

- **Planting:** Sow cover crops after harvesting main crops or during fallow periods. Allow them to grow and cover the soil.

- **Termination:** Before cover crops go to seed, terminate them by cutting or mowing, and leave the residue on the soil surface to decompose.

- **Benefits:** Cover crops add organic matter, fix nitrogen, improve soil structure, and suppress weeds.

## Composting in Place

Composting in place involves directly adding organic materials to the garden beds, where they decompose and enrich the soil.

- **Materials:** Use vegetable scraps, leaves, grass clippings, and other organic materials.

- **Application:** Dig shallow trenches or holes in the garden beds, add the organic materials, and cover them with soil or mulch.

- **Benefits:** Composting in place reduces waste, adds nutrients, and improves soil structure directly in the garden.

## Lasagna Gardening

Lasagna gardening, also known as sheet composting, is a no-till method that involves layering organic materials to build soil fertility.

- **Base Layer:** Start with a layer of cardboard or newspaper to suppress weeds.

- **Organic Layers:** Alternate layers of browns (e.g., straw, leaves) and greens (e.g., vegetable scraps, grass clippings). Aim for a final depth of 12-24 inches.

- **Planting:** Plant directly into the top layer or add a layer of finished compost or soil for planting.

- **Benefits:** Lasagna gardening improves soil fertility, retains moisture, and reduces weeds without tilling.

## Crop Rotation

Crop rotation involves growing different types of crops in succession in the same area to improve soil health and reduce pest and disease pressure.

- **Plan Rotations:** Group crops based on their botanical families and rotate them annually or biannually.

- **Diverse Crops:** Include a variety of crops with different root structures and nutrient requirements.

- **Benefits:** Crop rotation disrupts pest and disease cycles, reduces soil depletion, and promotes diverse soil biology.

## Implementing No-Till Gardening

Transitioning to no-till gardening involves several steps and considerations.

## Starting with No-Till

1. **Assess Soil Health:** Evaluate the current health and structure of your soil. Conduct a soil test to determine nutrient levels and pH.

2. **Plan for Cover Crops:** Select appropriate cover crops based

on your climate, soil type, and gardening goals. Plan for their integration into your cropping schedule.

3. **Mulch Application:** Begin applying organic mulch to your garden beds to protect the soil and add organic matter.

4. **Reduce Tillage Gradually:** If you are currently tilling, gradually reduce the frequency and intensity of tillage while increasing organic matter inputs.

## Ongoing Maintenance

- **Monitor Soil Health:** Regularly assess soil structure, moisture, and fertility. Adjust practices as needed based on observations and soil test results.

- **Manage Weeds:** Use mulch and cover crops to suppress weeds. Hand-weed or use organic herbicides if necessary.

- **Fertilize Organically:** Use compost, cover crops, and organic fertilizers to maintain soil fertility. Avoid synthetic fertilizers that can harm soil biology.

- **Maintain Biodiversity:** Grow a diverse range of crops and cover crops to promote a healthy, resilient soil ecosystem.

## Challenges and Solutions in No-Till Gardening

While no-till gardening offers many benefits, it also presents some challenges. Here are common challenges and solutions:

## Weed Management

Weeds can be more challenging to manage in no-till systems.

- **Solution:** Use thick layers of mulch and cover crops to suppress weeds. Hand-weed regularly and consider using organic mulch like cardboard or newspaper for added weed suppression.

## Soil Compaction

Soil compaction can occur if the soil is not managed properly.

- **Solution:** Avoid walking on garden beds and use pathways to reduce soil compaction. Incorporate organic matter to improve soil structure and prevent compaction.

## Pest and Disease Management

Pests and diseases can be more persistent without tillage to disrupt their life cycles.

- **Solution:** Practice crop rotation and use cover crops to break pest and disease cycles. Introduce beneficial insects and use organic pest control methods to manage pest populations.

## Initial Transition

The transition to no-till gardening can be challenging and may require adjustments.

- **Solution:** Start small and gradually expand your no-till practices.

Experiment with different cover crops, mulches, and composting methods to find what works best for your garden.

Raised beds and no-till gardening are powerful techniques that can transform your gardening experience. Raised beds offer numerous benefits, including improved soil quality, extended growing seasons, easier maintenance, and better pest control. Building and maintaining raised beds requires careful planning and regular upkeep to ensure long-term success.

No-till gardening, on the other hand, emphasizes soil health through minimal disturbance and the addition of organic matter. This method supports a thriving soil ecosystem, improves soil structure, and enhances fertility while reducing erosion and compaction. By implementing no-till techniques such as mulching, cover cropping, composting in place, and crop rotation, you can create a sustainable and productive garden.

# Chapter 5

# Growing Your Crops

Growing your own crops from seed can be a highly rewarding experience. It gives you control over the entire growing process, from choosing the right varieties to harvesting and saving seeds for the next season. This chapter will guide you through the essential aspects of seed starting and saving, selecting seeds, starting seeds indoors, and seed saving techniques.

## Seed Starting and Saving

Starting your own seeds and saving them for future use are fundamental skills for any homesteader. These practices allow you to grow a wide variety of crops, ensure the purity of your plant varieties, and save money.

## Benefits of Seed Starting

Starting your own seeds offers several advantages over purchasing transplants from a nursery.

1. **Variety Selection:** You have access to a broader range of plant varieties, including heirlooms and specialty crops that are not commonly available as transplants.

2. **Cost Savings:** Growing plants from seed is generally more economical than buying transplants.

3. **Control Over Growing Conditions:** You can provide optimal conditions for seed germination and early growth, leading to healthier plants.

4. **Season Extension:** Starting seeds indoors allows you to get a jump on the growing season, giving your plants a head start before transplanting them outdoors.

## Benefits of Seed Saving

Saving your own seeds can lead to greater self-sufficiency and resilience in your gardening practices.

1. **Cost Efficiency:** Saving seeds reduces the need to purchase new seeds each year.

2. **Adaptation:** Seeds saved from plants grown in your specific conditions will gradually adapt to your local environment, leading to improved performance over time.

3. **Preservation:** Saving seeds from heirloom varieties helps preserve genetic diversity and heritage crops.

4. **Self-Sufficiency:** Maintaining a personal seed bank ensures you always have seeds available, even in times of shortage or economic uncertainty.

## Selecting Seeds

Choosing the right seeds is the first step in successful crop production. The selection process involves considering factors such as climate, soil type, crop preferences, and intended use.

## Types of Seeds

There are several types of seeds to choose from, each with its own characteristics and benefits.

## Heirloom Seeds

Heirloom seeds are open-pollinated varieties that have been passed down through generations. They are prized for their unique flavors, colors, and historical significance.

- **Genetic Diversity:** Heirloom varieties contribute to biodiversity and can adapt well to local growing conditions.

- **Seed Saving:** Since they are open-pollinated, heirloom seeds can be saved and replanted each year, retaining their unique characteristics.

## Hybrid Seeds

Hybrid seeds are the result of cross-pollination between two different parent plants. They are bred to combine desirable traits such as disease resistance, high yield, and uniformity.

- **Vigor:** Hybrid seeds often exhibit hybrid vigor, resulting in strong, productive plants.

- **Reliability:** Hybrids are bred for specific traits, making them reliable performers in various conditions.

- **Seed Saving:** Seeds saved from hybrid plants may not produce true-to-type offspring, so it's generally best to purchase new hybrid seeds each year.

## Open-Pollinated Seeds

Open-pollinated seeds are those that are pollinated naturally by insects, birds, wind, or other natural mechanisms. They can produce true-to-type plants if properly isolated from other varieties.

- **Seed Saving:** Open-pollinated seeds can be saved and replanted, maintaining the same characteristics year after year.

- **Genetic Stability:** These seeds tend to be genetically stable and can adapt to local conditions over time.

## Genetically Modified Seeds (GMO)

Genetically modified seeds are developed using genetic engineering techniques to introduce specific traits, such as pest resistance or herbicide tolerance.

- **Controversy:** GMOs are subject to debate regarding their safety, environmental impact, and ethical considerations.

- **Regulation:** They are heavily regulated and may not be available to home gardeners.

## Factors to Consider When Selecting Seeds

When selecting seeds, consider the following factors to ensure the best results for your garden.

## Climate

Choose seeds that are well-suited to your local climate. Consider factors such as temperature ranges, growing season length, and precipitation patterns.

- **Hardiness Zones:** Check the USDA Plant Hardiness Zone Map to determine your zone and select varieties that will thrive in your area.

- **Heat and Drought Tolerance:** In hot or dry climates, look for varieties that are tolerant of heat and drought conditions.

## Soil Type

Different plants have different soil preferences. Consider your soil type and select seeds that will grow well in your garden's conditions.

- **Soil pH:** Most vegetables prefer a slightly acidic to neutral pH (6.0-7.0). Test your soil pH and choose varieties that match your soil conditions.

- **Soil Texture:** Sandy soils drain quickly, while clay soils retain moisture. Choose plants that are well-suited to your soil texture.

## Space and Layout

Consider the amount of space you have available and plan your garden layout accordingly.

- **Plant Size:** Select varieties that will fit your garden space and design. Compact or dwarf varieties are ideal for small gardens or containers.

- **Spacing Requirements:** Ensure you have enough space for proper plant spacing to avoid overcrowding and promote healthy growth.

## Disease and Pest Resistance

Choose varieties that are resistant to common diseases and pests in your area.

- **Disease-Resistant Varieties:** Look for varieties labeled as resistant to specific diseases such as blight, mildew, or rust.

- **Pest-Resistant Varieties:** Some varieties are bred for resistance to pests such as aphids, beetles, or nematodes.

## Intended Use

Select seeds based on how you intend to use the harvested crops.

- **Fresh Eating:** If you plan to eat your produce fresh, look for varieties known for their flavor and texture.

- **Preserving:** For canning, freezing, or drying, choose varieties that are well-suited for preservation.

- **Storage:** Some crops are better for long-term storage. Look for varieties labeled as good keepers or storage varieties.

## Starting Seeds Indoors

Starting seeds indoors allows you to control the early stages of plant growth, ensuring strong, healthy seedlings ready for transplanting. This section covers the essential steps and considerations for starting seeds indoors.

## Materials Needed

Gather the necessary materials before starting your seeds.

- **Seeds:** Select high-quality seeds suited to your growing conditions and preferences.

- **Containers:** Use seed trays, cell packs, pots, or recycled containers with drainage holes.

- **Seed Starting Mix:** Use a sterile, soilless seed starting mix to provide a light, well-draining medium for seed germination.

- **Labels:** Label each container with the seed variety and planting date.

- **Watering Can or Spray Bottle:** Use for gentle watering to avoid disturbing the seeds.

- **Grow Lights:** Provide sufficient light for seedlings if natural light is inadequate.

- **Heat Mat:** Optional, but beneficial for maintaining consistent soil temperatures for germination.

## Planting Seeds

Proper planting techniques ensure successful seed germination and healthy seedlings.

1. **Fill Containers:** Fill your containers with seed starting mix, leaving about half an inch of space at the top.

2. **Moisten Soil:** Lightly water the soil to moisten it evenly.

3. **Plant Seeds:** Follow the seed packet instructions for planting depth and spacing. Generally, plant seeds to a depth of 2-3 times their diameter.

4. **Cover Seeds:** Gently cover the seeds with soil mix and lightly tamp down to ensure good seed-to-soil contact.

5. **Label Containers:** Clearly label each container with the seed variety and planting date.

## Providing Optimal Conditions

Creating the right environment is crucial for seed germination and early growth.

## Light

Seedlings need plenty of light to grow strong and healthy.

- **Natural Light:** Place seedlings in a south-facing window with at least 6-8 hours of direct sunlight per day.

- **Artificial Light:** Use fluorescent or LED grow lights to supplement natural light. Position lights 2-4 inches above the seedlings and keep them on for 14-16 hours per day.

## Temperature

Maintain consistent temperatures to promote seed germination and growth.

- **Germination Temperature:** Most seeds germinate best at temperatures between 65-75°F. Use a heat mat if necessary to maintain soil temperature.

- **Growing Temperature:** After germination, maintain air temperatures between 60-70°F during the day and slightly cooler at night.

## Water

Proper watering is essential for seedling health.

- **Soil Moisture:** Keep the soil consistently moist but not waterlogged. Use a watering can or spray bottle to water gently.

- **Humidity:** Maintain moderate humidity levels to prevent seedlings from drying out. Use a humidity dome or cover with plastic wrap if necessary, but remove it once seedlings emerge to prevent damping-off disease.

# Caring for Seedlings

Once seeds have germinated, provide the necessary care to ensure healthy growth.

# Thinning

Thin seedlings to avoid overcrowding and competition for resources.

- **Timing:** Thin seedlings when they develop their first true leaves.

- **Method:** Use scissors to snip excess seedlings at soil level, leaving the strongest seedling in each cell or pot.

# Fertilizing

Provide nutrients to support seedling growth.

- **Timing:** Begin fertilizing when seedlings develop their first true leaves.

- **Fertilizer:** Use a diluted, balanced liquid fertilizer (e.g., 10-10-10) or fish emulsion at half-strength. Apply every 1-2 weeks.

# Hardening Off

Prepare seedlings for the transition to outdoor conditions through a process called hardening off.

- **Timing:** Start hardening off seedlings about 1-2 weeks before transplanting.

- **Process:** Gradually expose seedlings to outdoor conditions by placing them in a sheltered location for a few hours each day. Increase exposure time and reduce protection over several days.

- **Protection:** Protect seedlings from strong winds, direct sunlight, and temperature extremes during the hardening-off period.

## Transplanting Seedlings

Transplanting seedlings into the garden requires careful preparation and execution.

1. **Choose the Right Time:** Transplant seedlings after the last frost date when the soil has warmed and nighttime temperatures are consistently above 50°F.

2. **Prepare the Soil:** Amend the garden soil with compost or other organic matter to improve fertility and drainage.

3. **Water Seedlings:** Water seedlings thoroughly before transplanting to reduce transplant shock.

4. **Planting Holes:** Dig planting holes slightly larger than the seedling root balls.

5. **Transplant Seedlings:** Gently remove seedlings from their containers, handling them by the leaves, not the stems. Place seedlings in the planting holes and firm the soil around them.

6. **Water and Mulch:** Water newly transplanted seedlings and apply mulch around the base to retain moisture and suppress weeds.

## Seed Saving Techniques

Saving your own seeds allows you to preserve plant varieties and develop a self-sufficient gardening practice. This section covers essential seed-saving techniques, including selecting plants, harvesting seeds, and proper storage.

## Selecting Plants for Seed Saving

Choosing the right plants for seed saving is crucial for maintaining genetic purity and desirable traits.

## Open-Pollinated Varieties

Save seeds from open-pollinated varieties to ensure that the offspring will resemble the parent plants.

- **Heirloom Varieties:** Heirloom varieties are ideal for seed saving as they are open-pollinated and have been maintained for their unique traits.

## Avoid Hybrid Varieties

Avoid saving seeds from hybrid varieties, as they may not produce true-to-type offspring and can exhibit unpredictable traits.

## Select Healthy Plants

Choose healthy, vigorous plants free from disease and pests for seed saving. Select plants that exhibit desirable traits such as high yield, disease resistance, and good flavor.

## Isolation Techniques

Prevent cross-pollination between different varieties to maintain genetic purity.

## Physical Barriers

Use physical barriers to isolate plants and prevent cross-pollination.

- **Cages and Tents:** Cover plants with cages or tents made from fine mesh or fabric to exclude pollinators.

- **Distance:** Plant different varieties of the same species at a sufficient distance from each other to reduce the risk of cross-pollination. The required distance varies by species.

## Hand Pollination

Hand-pollinate flowers to control pollination and ensure genetic purity.

- **Method:** Use a small brush or cotton swab to transfer pollen from the male flower (staminate) to the female flower (pistillate).

- **Isolation:** Isolate hand-pollinated flowers by covering them with a fine mesh bag or cloth to prevent unwanted pollination.

## Harvesting Seeds

Harvesting seeds at the right time ensures their viability and quality.

## Dry Seeds

Dry seeds are harvested from the seed pods, heads, or capsules of plants. Common dry seeds include beans, peas, lettuce, and flowers.

- **Maturity:** Allow seeds to mature fully on the plant. The seed pods or heads should be dry and brittle.

- **Harvesting:** Collect dry seed pods or heads and place them in a paper bag or envelope. Allow them to dry further in a well-ventilated area.

- **Threshing:** Gently crush the seed pods or heads to release the seeds. Use a screen or sieve to separate the seeds from the chaff.

## Wet Seeds

Wet seeds are harvested from the fleshy fruits of plants. Common wet seeds include tomatoes, cucumbers, melons, and squash.

- **Maturity:** Allow the fruit to ripen fully on the plant.

- **Harvesting:** Remove the seeds from the fruit and rinse them to remove any pulp or residue.

- **Fermentation (for some seeds):** Some seeds, like tomatoes, benefit from a brief fermentation process to remove the gel coating. Place the seeds in a jar with a small amount of water and let them ferment for 2-3 days. Rinse and dry the seeds afterward.

## Cleaning and Drying Seeds

Proper cleaning and drying are essential for seed viability and storage.

## Cleaning

Remove any remaining debris, pulp, or chaff from the seeds.

- **Dry Seeds:** Use screens, sieves, or fans to winnow away debris from dry seeds.

- **Wet Seeds:** Rinse wet seeds thoroughly to remove any pulp or residue.

## Drying

Dry seeds thoroughly before storage to prevent mold and decay.

- **Spread Seeds:** Spread seeds in a single layer on a screen, paper towel, or cloth.

- **Drying Location:** Place seeds in a well-ventilated area out of direct sunlight. Use a fan to improve air circulation if necessary.

- **Drying Time:** Allow seeds to dry for several days to a few weeks, depending on the seed size and moisture content. Seeds should be hard and brittle when fully dry.

## Storing Seeds

Proper storage is essential to maintain seed viability and longevity.

## Storage Containers

Use airtight containers to protect seeds from moisture, pests, and temperature fluctuations.

- **Glass Jars:** Glass jars with tight-fitting lids are ideal for seed storage.

- **Plastic Containers:** Use plastic containers with airtight seals for long-term storage.

- **Paper Envelopes:** Store seeds in paper envelopes for short-term storage or if moisture control is not a concern.

## Storage Conditions

Store seeds in a cool, dry, and dark location to preserve their viability.

- **Temperature:** Maintain a consistent storage temperature between 32-41°F (0-5°C). A refrigerator is an ideal storage location.

- **Humidity:** Keep seeds dry by adding a desiccant, such as silica gel packets, to the storage container.

- **Darkness:** Store seeds in opaque containers or a dark location to protect them from light exposure.

## Labeling

Clearly label each container with the seed variety, harvest date, and any other relevant information.

- **Variety Name:** Include the common name and variety name of the seeds.

- **Harvest Date:** Record the date the seeds were harvested to track their age and viability.

## Seed Viability and Germination Testing

Testing seed viability and germination rates ensures that your saved seeds are still capable of producing healthy plants.

## Viability Testing

Perform a simple viability test to check if seeds are still viable.

- **Float Test:** Place seeds in a bowl of water. Viable seeds will sink, while non-viable seeds will float. Note that this test is not always accurate for all seed types.

## Germination Testing

Conduct a germination test to determine the percentage of seeds that will germinate.

- **Method:** Place a sample of seeds (e.g., 10 or 20 seeds) between moist paper towels or in a seed starting tray with a light soil mix. Keep the seeds warm and moist.

- **Observation:** Monitor the seeds for germination over the expected germination period (usually 7-14 days). Count the number of seeds that germinate and calculate the germination rate as a percentage.

- **Adjust Planting Rate:** Use the germination rate to adjust your planting rate. If the germination rate is low, plant more seeds to achieve the desired number of seedlings.

## Advanced Seed Saving Techniques

For experienced gardeners, advanced seed saving techniques can help maintain the genetic purity and health of your saved seeds.

## Isolation by Time

Isolate plants by timing their flowering periods to prevent cross-pollination.

- **Staggered Planting:** Plant different varieties at different times so that their flowering periods do not overlap.

## Rogueing

Rogueing involves removing undesirable plants from the population to maintain genetic purity and desirable traits.

- **Identify Traits:** Identify and remove plants that do not exhibit desired traits, such as poor growth, disease susceptibility, or off-type characteristics.

- **Timing:** Perform rogueing early in the growing season before plants flower and produce seeds.

## Population Size

Maintain a sufficient population size to preserve genetic diversity and avoid inbreeding depression.

- **Minimum Number:** For most plants, maintain a minimum of 20-50 plants for seed saving. Some plants, such as corn, require larger populations to maintain genetic diversity.

## Crop-Specific Seed Saving Tips

Different crops have specific requirements and techniques for successful seed saving. Here are tips for some common garden crops:

### Tomatoes

- **Fermentation:** Ferment tomato seeds to remove the gel coating that inhibits germination.

- **Isolation:** Tomatoes are self-pollinating but can cross-pollinate with other varieties. Isolate by distance or use physical barriers.

### Peppers

- **Isolation:** Peppers can cross-pollinate with other varieties. Isolate by distance or use physical barriers.

- **Harvesting:** Allow peppers to fully ripen and change color before harvesting seeds.

### Beans and Peas

- **Drying:** Allow beans and peas to dry fully on the plant before harvesting.

- **Threshing:** Gently crush the pods to release the seeds. Use a screen to separate seeds from debris.

## Cucumbers and Melons

- **Fermentation:** Ferment cucumber seeds to remove the gel coating. Melon seeds do not require fermentation.

- **Isolation:** Cucumbers and melons can cross-pollinate with other varieties. Isolate by distance or use physical barriers.

## Squash and Pumpkins

- **Isolation:** Squash and pumpkins can cross-pollinate with other varieties. Isolate by distance or use physical barriers.

- **Harvesting:** Allow fruits to fully mature on the plant before harvesting seeds.

## Lettuce

- **Drying:** Allow lettuce plants to bolt and produce seed stalks. Harvest seed heads when they are dry and brittle.

- **Threshing:** Gently crush the seed heads to release the seeds. Use a screen to separate seeds from debris.

Chapter 3 covered essential skills for homesteaders and gardening enthusiasts, focusing on growing crops from seed, starting seeds indoors, and saving seeds. We explored the benefits of seed starting, such as variety selection, cost savings, and season extension, along with the advantages of seed saving, like cost efficiency and self-sufficiency.

Choosing the right seeds involves considering climate, soil type, space, disease resistance, and intended use. We discussed different seed types like heirloom, hybrid, open-pollinated, and genetically modified seeds, highlighting their characteristics and selection factors.

Starting seeds indoors requires careful planning and maintenance of optimal conditions for healthy seedling growth. Steps included planting, providing light, temperature, water, and nutrients, along with transplanting and post-transplant care.

# Chapter 6

# Planting and Transplanting

Planting and transplanting are fundamental processes in gardening that significantly affect plant growth and yield. This section will explore the differences between direct sowing and transplanting, proper planting techniques, and the practice of companion planting to enhance plant health and productivity.

## Direct Sowing vs. Transplanting

Choosing between direct sowing and transplanting depends on several factors, including the type of crop, climate, and available resources. Each method has its own set of advantages and disadvantages.

## Direct Sowing

Direct sowing involves planting seeds directly into the garden soil where they will grow to maturity. This method is often used for crops that do not transplant well or have a short growing season.

## Advantages of Direct Sowing

**Less Labor:** Direct sowing eliminates the need for handling and caring for seedlings indoors.

**Stronger Root Systems:** Plants develop robust root systems in their final location without the shock of transplanting.

**Natural Selection:** Seeds germinate under natural conditions, resulting in stronger and more resilient plants.

## Disadvantages of Direct Sowing

**Slower Start:** Seeds may take longer to germinate outdoors due to fluctuating soil temperatures and moisture levels.

**Vulnerability:** Seeds and young seedlings are more vulnerable to pests, diseases, and adverse weather conditions.

**Weed Competition:** Seeds may have to compete with weeds for nutrients, water, and light.

## Ideal Crops for Direct Sowing

- **Root Crops:** Carrots, radishes, beets, and turnips.

- **Legumes:** Beans and peas.

- **Squash Family:** Cucumbers, melons, pumpkins, and zucchini.

- **Leafy Greens:** Lettuce, spinach, arugula, and chard.

# Transplanting

Transplanting involves starting seeds indoors and then moving the seedlings to the garden once they are strong enough. This method is beneficial for crops that require a longer growing season or need protection during their early growth stages.

## Advantages of Transplanting

**Early Start:** Transplants get a head start on the growing season, allowing for earlier harvests.

**Controlled Environment:** Seedlings are protected from pests, diseases, and harsh weather conditions.

**Efficient Use of Space:** Garden space is optimized by planting strong seedlings that have a higher survival rate.

## Disadvantages of Transplanting

**Labor Intensive:** Requires additional work to start seeds indoors and care for seedlings.

**Transplant Shock:** Seedlings may experience transplant shock, which can temporarily slow growth.

**Resource Intensive:** Requires additional materials such as seed starting mix, containers, and grow lights.

## Ideal Crops for Transplanting

- **Nightshades:** Tomatoes, peppers, and eggplants.

- **Brassicas:** Broccoli, cauliflower, cabbage, and kale.

- **Herbs:** Basil, parsley, and thyme.

- **Flowers:** Marigolds, zinnias, and petunias.

## Proper Planting Techniques

Whether direct sowing or transplanting, using proper planting techniques is crucial for ensuring healthy plant growth and maximizing yield.

## Direct Sowing Techniques

### Soil Preparation

**Clear the Area:** Remove any weeds, rocks, and debris from the planting area.

**Loosen the Soil:** Use a garden fork or tiller to loosen the soil to a depth of 6-8 inches.

**Amend the Soil:** Incorporate organic matter such as compost or well-rotted manure to improve soil fertility and structure.

### Sowing Seeds

**Mark Rows:** Use a string or garden marker to create straight rows for planting.

**Planting Depth:** Follow seed packet instructions for the correct planting depth. A general rule is to plant seeds at a depth of 2-3 times their diameter.

**Spacing:** Space seeds according to the recommended spacing on the seed packet to avoid overcrowding.

**Cover Seeds:** Lightly cover seeds with soil and gently firm the soil to ensure good seed-to-soil contact.

**Watering:** Water the area gently but thoroughly to moisten the soil without washing away the seeds.

## Thinning Seedlings

**Timing:** Thin seedlings once they have developed their first true leaves.

**Spacing:** Remove excess seedlings to provide adequate space for remaining plants, following the recommended spacing on the seed packet.

**Method:** Use scissors to snip seedlings at soil level to avoid disturbing the roots of neighboring plants.

## Transplanting Techniques

### Preparing Seedlings

**Hardening Off:** Gradually acclimate seedlings to outdoor conditions by placing them outside for a few hours each day, increasing the exposure time over 7-10 days.

**Watering:** Water seedlings thoroughly a few hours before transplanting to reduce transplant shock.

### Preparing the Planting Area

1. **Soil Preparation:** Follow the same soil preparation steps as for direct sowing.

2. **Planting Holes:** Dig planting holes slightly larger than the seedling root balls.

## Transplanting Seedlings

**Removing Seedlings:** Gently remove seedlings from their containers, handling them by the leaves to avoid damaging the stems.

**Planting Depth:** Plant seedlings at the same depth they were growing in their containers, except for tomatoes, which can be planted deeper to encourage additional root growth.

**Firming Soil:** Firm the soil around the seedlings to eliminate air pockets and provide stability.

**Watering:** Water seedlings immediately after transplanting to help settle the soil and reduce transplant shock.

**Mulching:** Apply a layer of mulch around the base of the seedlings to retain moisture and suppress weeds.

## Companion Planting for Success

Companion planting involves growing different plants together to benefit each other in various ways. This practice can improve plant health, enhance growth, deter pests, and increase yields.

## Benefits of Companion Planting

**Pest Control:** Certain plants repel pests or attract beneficial insects that prey on pests.

**Disease Prevention:** Some plants can reduce the spread of diseases by creating physical barriers or through allelopathy (the release of chemicals that inhibit the growth of other plants).

**Improved Growth:** Companion plants can provide shade, support, or improve soil fertility through nitrogen fixation.

**Space Efficiency:** Growing compatible plants together maximizes garden space and increases biodiversity.

## Common Companion Planting Combinations

### Three Sisters

The Three Sisters is a traditional Native American planting technique that involves growing corn, beans, and squash together.

- **Corn:** Provides a natural trellis for the beans to climb.

- **Beans:** Fix nitrogen in the soil, benefiting the corn and squash.

- **Squash:** Spreads along the ground, suppressing weeds and retaining soil moisture.

### Tomatoes and Basil

Tomatoes and basil are classic companions that benefit each other in several ways.

- **Basil:** Repels pests such as aphids, whiteflies, and tomato hornworms. It may also improve the flavor of tomatoes.

## Carrots and Onions

Carrots and onions are effective companions for pest control.

- **Onions:** Repel carrot flies, while carrots help deter onion flies.

## Lettuce and Radishes

Lettuce and radishes grow well together and can be harvested at different times.

- **Radishes:** Mature quickly, making room for lettuce as they are harvested. Radish leaves can also provide some shade for young lettuce plants.

## Marigolds and Almost Everything

Marigolds are known for their pest-repelling properties and can be planted with many vegetables.

- **Marigolds:** Repel nematodes, aphids, and other pests. They also attract beneficial insects like ladybugs.

## Designing a Companion Planting Garden

Designing a companion planting garden involves careful planning and understanding the relationships between different plants.

## Mapping the Garden

1. **Create a Garden Plan:** Draw a map of your garden beds, noting the location and spacing of each plant.

2. **Group Plants by Companions:** Group compatible plants together, considering their growth habits, spacing needs, and maturity times.

## Succession Planting

1. **Plan for Continuous Harvest:** Use succession planting techniques to ensure a continuous harvest throughout the growing season.

2. **Rotate Crops:** Rotate crops each season to prevent the buildup of pests and diseases.

## Interplanting

1. **Maximize Space:** Interplant fast-growing crops with slower-growing ones to make the most of available space.

2. **Improve Biodiversity:** Mix different types of plants to enhance biodiversity and reduce the risk of pests and diseases.

## Practical Tips for Successful Planting and Transplanting

Here are some additional practical tips to ensure successful planting and transplanting in your garden.

## Timing

1. **Know Your Frost Dates:** Be aware of your local frost dates and plan your planting schedule accordingly.

2. **Seasonal Considerations:** Plant cool-season crops in early spring and fall, and warm-season crops after the last frost date.

## Soil Health

1. **Test Your Soil:** Conduct a soil test to determine nutrient levels and pH. Amend the soil as needed to create optimal growing conditions.

2. **Maintain Soil Fertility:** Regularly add organic matter such as compost or manure to maintain soil fertility and structure.

## Watering

1. **Consistent Moisture:** Keep the soil consistently moist, especially during seed germination and seedling establishment.

2. **Avoid Overwatering:** Avoid overwatering, which can lead to root rot and other issues.

## Pest and Disease Management

1. **Monitor Regularly:** Regularly inspect plants for signs of pests and diseases. Early detection and intervention are key to managing problems.

2. **Use Integrated Pest Management (IPM):** Employ a combination

of cultural, biological, and organic control methods to manage pests and diseases.

## Mulching

1. **Conserve Moisture:** Apply mulch to conserve soil moisture, suppress weeds, and improve soil health.

2. **Organic Mulch:** Use organic mulch such as straw, leaves, or grass clippings to add nutrients to the soil as it decomposes.

## Record Keeping

1. **Garden Journal:** Keep a garden journal to record planting dates, varieties, weather conditions, and any issues that arise.

2. **Evaluate and Adjust:** Use your records to evaluate the success of different planting techniques and make adjustments for future seasons.

Planting and transplanting are critical steps in the gardening process that require careful planning and execution. Whether you choose direct sowing or transplanting, proper planting techniques will help ensure the health and productivity of your crops. Additionally, companion planting can enhance plant growth, deter pests, and improve garden biodiversity.

# Chapter 7

# Crop Rotation and Succession Planting

Crop rotation and succession planting are essential techniques for maximizing garden productivity, maintaining soil health, and reducing pest and disease pressures. This section will delve into the importance of crop rotation, how to plan a rotation schedule, and the methods of succession planting for a continuous harvest.

## Importance of Crop Rotation

Crop rotation is the practice of growing different types of crops in the same area across different seasons or years. This practice has been used for centuries to enhance soil fertility and manage pests and diseases.

## Soil Fertility and Structure

Rotating crops helps maintain and improve soil fertility and structure. Different plants have varying nutrient needs and rooting depths, which can help balance soil nutrient levels and prevent depletion.

- **Nutrient Management:** Some crops, like legumes, can fix atmospheric nitrogen into the soil, enriching it for subsequent crops. Rotating these with heavy feeders like corn or tomatoes ensures a balanced nutrient profile.

- **Soil Structure:** Deep-rooted crops like carrots or parsnips can break up compacted soil layers, improving aeration and drainage for subsequent crops with shallower roots.

## Pest and Disease Control

Crop rotation helps break the life cycles of pests and diseases. Many pests and pathogens are host-specific, meaning they only affect certain types of plants. By rotating crops, you can disrupt their life cycles and reduce their populations.

- **Pest Management:** Rotating crops prevents pests from establishing permanent populations in the soil. For example, rotating brassicas (cabbage, broccoli) with unrelated crops can help control cabbage root maggot populations.

- **Disease Prevention:** Soil-borne diseases like verticillium wilt or clubroot can persist in the soil for years. Rotating susceptible crops with resistant ones can reduce disease pressure.

## Weed Management

Different crops can suppress weeds in various ways. Rotating crops with different growth habits, canopy structures, and planting times can reduce weed pressure.

- **Canopy Cover:** Crops that form a dense canopy, such as squash or pumpkins, can shade out weeds, reducing their growth and seed production.

- **Allelopathy:** Some crops, like rye, release chemicals into the soil that inhibit weed seed germination.

## Planning a Rotation Schedule

Effective crop rotation requires careful planning to ensure that crops are rotated in a way that maximizes their benefits. This involves understanding the plant families, nutrient needs, and potential pests and diseases of each crop.

## Understanding Plant Families

Grouping crops by their botanical families helps simplify the planning process. Plants within the same family often share similar pest and disease susceptibilities and nutrient needs.

- **Nightshade Family (Solanaceae):** Tomatoes, peppers, eggplants, potatoes

- **Brassica Family (Brassicaceae):** Cabbage, broccoli, cauliflower, kale, radishes

- **Legume Family (Fabaceae):** Beans, peas, lentils, peanuts

- **Cucurbit Family (Cucurbitaceae):** Cucumbers, melons, squash, pumpkins

- **Onion Family (Alliaceae):** Onions, garlic, leeks, shallots

- **Carrot Family (Apiaceae):** Carrots, parsley, celery, dill

## Determining Rotation Groups

Create rotation groups based on the plant families and their specific requirements and benefits. Each group should be rotated through different garden sections each year.

## Example Rotation Groups

1. **Group 1: Legumes**

   - Benefits: Fix nitrogen, improve soil fertility.

   - Crops: Beans, peas.

2. **Group 2: Heavy Feeders**

   - Benefits: High nutrient demand, follow nitrogen-fixing crops.

   - Crops: Corn, tomatoes, peppers, squash.

3. **Group 3: Light Feeders**

   - Benefits: Lower nutrient demand, follow heavy feeders.

   - Crops: Carrots, beets, onions.

4. **Group 4: Soil Builders**

   - Benefits: Improve soil structure and organic matter.

○ Crops: Cover crops (clover, rye), green manures.

## Developing a Rotation Plan

A rotation plan should cover multiple years and ensure that no crop family is grown in the same area consecutively. A common rotation cycle lasts three to four years.

## Example 4-Year Rotation Plan

- **Year 1:**

  ○ Bed 1: Legumes

  ○ Bed 2: Heavy Feeders

  ○ Bed 3: Light Feeders

  ○ Bed 4: Soil Builders

- **Year 2:**

  ○ Bed 1: Heavy Feeders

  ○ Bed 2: Light Feeders

  ○ Bed 3: Soil Builders

  ○ Bed 4: Legumes

- **Year 3:**

  ○ Bed 1: Light Feeders

- Bed 2: Soil Builders

- Bed 3: Legumes

- Bed 4: Heavy Feeders

- **Year 4:**

  - Bed 1: Soil Builders

  - Bed 2: Legumes

  - Bed 3: Heavy Feeders

  - Bed 4: Light Feeders

## Adjusting for Small Gardens

In small gardens, traditional crop rotation can be challenging due to space constraints. Here are some tips for adapting crop rotation principles to smaller spaces:

1. **Interplanting:** Grow a mix of crops in the same bed, ensuring they belong to different families and have complementary growth habits.

2. **Container Gardening:** Use containers to grow some crops, making it easier to rotate them each season.

3. **Vertical Gardening:** Utilize vertical space to grow crops like beans, cucumbers, and tomatoes, allowing for better rotation on the ground level.

# Succession Planting for Continuous Harvest

Succession planting is the practice of planting crops in intervals throughout the growing season to ensure a continuous harvest. This method maximizes garden productivity and provides a steady supply of fresh produce.

## Types of Succession Planting

There are several types of succession planting, each suited to different crops and growing conditions.

## Staggered Planting

Staggered planting involves sowing seeds of the same crop at regular intervals. This method ensures that plants mature at different times, providing a continuous harvest.

- **Example:** Sow lettuce seeds every two weeks to maintain a steady supply of fresh greens.

## Relay Planting

Relay planting involves planting a new crop in the space of a maturing crop before it is fully harvested. This method ensures that there is always something growing in the garden.

- **Example:** Plant bush beans between rows of maturing spinach. Once the spinach is harvested, the beans will take over the space.

## Intercropping

Intercropping involves growing two or more crops in the same space at the same time. The crops should have different growth habits and maturity rates to avoid competition and make efficient use of space.

- **Example:** Plant radishes and carrots together. The radishes will mature and be harvested before the carrots need the extra space.

## Multiple Cropping

Multiple cropping involves growing different crops in the same space in succession within a single growing season. This method maximizes the use of garden space and takes advantage of different growing periods.

- **Example:** Plant early spring crops like peas and spinach, followed by summer crops like tomatoes and peppers, and finish with fall crops like kale and turnips.

## Planning for Succession Planting

Effective succession planting requires careful planning and consideration of various factors, including crop maturity times, climate, and garden space.

## Understanding Crop Maturity Times

Different crops have varying maturity times, which affects their suitability for succession planting. Knowing the days to maturity for each crop helps plan the planting schedule.

- **Fast Maturing:** Radishes (20-30 days), lettuce (30-60 days), spinach (40-50 days)

- **Medium Maturing:** Carrots (60-80 days), beans (50-70 days), beets (50-60 days)

- **Slow Maturing:** Tomatoes (70-100 days), peppers (70-100 days), squash (70-90 days)

## Considering Climate and Frost Dates

Your local climate and frost dates determine the length of the growing season and influence succession planting strategies.

- **Frost-Free Period:** Calculate the number of frost-free days in your growing season to plan multiple plantings.

- **Season Extension:** Use season extension techniques such as row covers, cold frames, or greenhouses to extend the growing season and allow for more succession plantings.

## Garden Layout and Space Management

Efficient use of garden space is crucial for successful succession planting. Plan your garden layout to accommodate multiple plantings and ensure adequate space for each crop.

- **Row Spacing:** Use wider rows to allow for interplanting and relay planting.

- **Pathways:** Leave enough space between beds for easy access and maintenance.

## Practical Tips for Succession Planting

Here are some practical tips to help you implement succession planting in your garden.

## Pre-Sprouting Seeds

Pre-sprouting seeds before planting can give them a head start and ensure faster germination.

- **Method:** Soak seeds in water for 24 hours, then place them between damp paper towels until they sprout. Plant the pre-sprouted seeds in the garden.

## Using Transplants

Starting some crops indoors and transplanting them into the garden can help maintain a continuous harvest.

- **Example:** Start lettuce or kale indoors and transplant them into the garden when the previous crop is harvested.

## Efficient Watering

Efficient watering practices ensure that all plants receive adequate moisture, especially during the transition between successive plantings.

- **Drip Irrigation:** Use drip irrigation to provide consistent moisture to plants without wasting water.

- **Mulching:** Mulch around plants to retain moisture and reduce the need for frequent watering.

## Soil Fertility Management

Maintaining soil fertility is crucial for supporting multiple crops in succession. Regularly amend the soil with organic matter to replenish nutrients.

- **Composting:** Add compost to the soil between plantings to provide essential nutrients and improve soil structure.

- **Cover Crops:** Grow cover crops during fallow periods to fix nitrogen and improve soil fertility for future plantings.

## Monitoring and Adjusting

Regularly monitor the progress of your crops and adjust your succession planting plan as needed.

- **Observation:** Keep track of growth rates, harvest times, and any pest or disease issues.

- **Flexibility:** Be prepared to adjust planting schedules based on weather conditions and crop performance.

## Combining Crop Rotation and Succession Planting

Combining crop rotation and succession planting can further enhance garden productivity and soil health. Here's how to integrate these practices effectively.

## Rotational Succession Planting

Implement rotational succession planting by rotating crop families in succession throughout the growing season.

- **Example:** Start with early-season peas (legumes), followed by mid-season carrots (light feeders), and finish with late-season spinach (soil builders).

## Interplanting and Rotating

Combine interplanting and rotation to maximize space and benefits.

- **Example:** Interplant fast-growing radishes with slower-growing tomatoes. Once the radishes are harvested, plant a quick crop of lettuce in their place.

## Year-Round Planning

Plan for year-round gardening by incorporating both crop rotation and succession planting into your garden calendar.

- **Winter Planning:** Use winter months to plan crop rotations and succession schedules for the upcoming year.

- **Seasonal Adjustments:** Adjust plans based on seasonal changes and specific garden needs.

## Record Keeping and Analysis

Maintain detailed records of your crop rotations and succession plantings to evaluate their effectiveness and make improvements.

- **Garden Journal:** Record planting dates, crop varieties, harvest times, and any issues encountered.

- **Analysis:** Review your records at the end of each season to assess what worked well and what needs adjustment.

## Example of a Combined Crop Rotation and Succession Planting Plan

Here is an example of a combined crop rotation and succession planting plan for a small garden with four beds.

### Year 1

- **Bed 1:**

  - Spring: Peas (legumes)

  - Summer: Carrots (light feeders)

  - Fall: Spinach (soil builders)

- **Bed 2:**

  - Spring: Lettuce (light feeders)

  - Summer: Tomatoes (heavy feeders)

  - Fall: Cover Crop (clover)

- **Bed 3:**

- Spring: Radishes (light feeders)

  - Summer: Beans (legumes)

  - Fall: Kale (soil builders)

- **Bed 4:**

  - Spring: Garlic (light feeders)

  - Summer: Squash (heavy feeders)

  - Fall: Turnips (light feeders)

## Year 2

- **Bed 1:**

  - Spring: Lettuce (light feeders)

  - Summer: Tomatoes (heavy feeders)

  - Fall: Cover Crop (rye)

- **Bed 2:**

  - Spring: Radishes (light feeders)

  - Summer: Beans (legumes)

  - Fall: Kale (soil builders)

- **Bed 3:**

- Spring: Garlic (light feeders)

- Summer: Squash (heavy feeders)

- Fall: Turnips (light feeders)

- **Bed 4:**

  - Spring: Peas (legumes)

  - Summer: Carrots (light feeders)

  - Fall: Spinach (soil builders)

## Year 3

- **Bed 1:**

  - Spring: Radishes (light feeders)

  - Summer: Beans (legumes)

  - Fall: Kale (soil builders)

- **Bed 2:**

  - Spring: Garlic (light feeders)

  - Summer: Squash (heavy feeders)

  - Fall: Turnips (light feeders)

- **Bed 3:**

- Spring: Peas (legumes)

- Summer: Carrots (light feeders)

- Fall: Spinach (soil builders)

- **Bed 4:**

  - Spring: Lettuce (light feeders)

  - Summer: Tomatoes (heavy feeders)

  - Fall: Cover Crop (clover)

## Year 4

- **Bed 1:**

  - Spring: Garlic (light feeders)

  - Summer: Squash (heavy feeders)

  - Fall: Turnips (light feeders)

- **Bed 2:**

  - Spring: Peas (legumes)

  - Summer: Carrots (light feeders)

  - Fall: Spinach (soil builders)

- **Bed 3:**

- Spring: Lettuce (light feeders)

- Summer: Tomatoes (heavy feeders)

- Fall: Cover Crop (clover)

- **Bed 4:**

    - Spring: Radishes (light feeders)

    - Summer: Beans (legumes)

    - Fall: Kale (soil builders)

Crop rotation and succession planting are integral practices for maintaining a productive and sustainable garden. Crop rotation helps manage soil fertility, pests, diseases, and weeds, while succession planting ensures a continuous harvest throughout the growing season. By understanding the importance of these techniques, planning a detailed rotation schedule, and implementing effective succession planting strategies, you can maximize your garden's potential and enjoy a diverse, bountiful harvest year after year.

# Chapter 8

# Pest and Disease Management

Managing pests and diseases is a critical aspect of maintaining a healthy and productive garden. Integrated Pest Management (IPM) offers a holistic approach to pest control, combining various strategies to minimize damage and promote ecological balance. This chapter will cover the principles of IPM, how to identify common pests, organic pest control methods, and the role of beneficial insects and natural predators in the garden.

## Integrated Pest Management (IPM)

Integrated Pest Management (IPM) is an environmentally friendly approach to pest control that emphasizes the use of multiple strategies to manage pests in a sustainable manner. IPM focuses on long-term prevention and control of pests through a combination of biological, cultural, physical, and chemical methods.

## Principles of IPM

1. **Prevention:** The first line of defense in IPM is to prevent pest problems before they occur. This involves creating an environment that is less conducive to pest development.

2. **Monitoring:** Regularly inspect your garden to identify pest problems early. Use traps, visual inspections, and other monitoring techniques to keep track of pest populations.

3. **Identification:** Correctly identify pests and understand their life cycles to determine the best control methods.

4. **Thresholds:** Establish action thresholds for different pests. These thresholds are the levels at which pest populations or damage become unacceptable and require intervention.

5. **Control Methods:** Implement a combination of control methods, including cultural, biological, physical, and chemical controls, to manage pest populations.

6. **Evaluation:** Continuously evaluate the effectiveness of your pest management strategies and make adjustments as needed.

## Steps in Implementing IPM

## Step 1: Prevention

Prevention is the foundation of IPM and involves practices that minimize the likelihood of pest problems.

- **Crop Rotation:** Rotate crops to prevent the buildup of pests and diseases associated with specific plants.

- **Diverse Planting:** Plant a diverse range of crops to reduce the risk of widespread pest infestations.

- **Resistant Varieties:** Choose pest-resistant plant varieties to reduce susceptibility to pests and diseases.

- **Sanitation:** Keep the garden clean by removing plant debris, weeds, and other materials that can harbor pests.

- **Healthy Soil:** Maintain healthy soil through proper fertilization, irrigation, and organic matter additions to support strong plant growth.

## Step 2: Monitoring

Regular monitoring helps detect pest problems early and assess the effectiveness of control measures.

- **Visual Inspections:** Regularly inspect plants for signs of pests, such as chewed leaves, discoloration, or stunted growth.

- **Traps:** Use traps, such as sticky traps or pheromone traps, to monitor pest populations.

- **Records:** Keep detailed records of pest observations, including the type of pest, population levels, and any damage observed.

## Step 3: Identification

Accurate identification of pests is crucial for selecting the most effective control methods.

- **Reference Materials:** Use field guides, online resources, or consult with local extension services to identify pests.

- **Life Cycles:** Understand the life cycles of pests to target them at the most vulnerable stages.

## Step 4: Establishing Thresholds

Determine the action thresholds for different pests based on the level of damage that can be tolerated before intervention is necessary.

- **Economic Thresholds:** For crops grown for sale, the economic threshold is the pest population level at which the cost of damage exceeds the cost of control.

- **Aesthetic Thresholds:** For ornamental plants, the threshold is based on the level of damage that affects the appearance of the plants.

## Step 5: Control Methods

Use a combination of control methods to manage pest populations effectively and sustainably.

- **Cultural Controls:** Modify cultural practices to reduce pest populations, such as adjusting planting times, spacing, and irrigation practices.

- **Biological Controls:** Introduce or enhance populations of natural predators, parasitoids, and pathogens that target pests.

- **Physical Controls:** Use barriers, traps, and manual removal to reduce pest populations.

- **Chemical Controls:** Use pesticides as a last resort and choose products that are least harmful to non-target organisms and the environment.

## Step 6: Evaluation

Evaluate the effectiveness of your pest management strategies and adjust them as needed.

- **Effectiveness:** Assess whether pest populations and damage levels have been reduced to acceptable levels.

- **Side Effects:** Monitor for any unintended side effects on non-target organisms or the environment.

- **Adaptation:** Make changes to your IPM plan based on your evaluations and new information.

## Identifying Common Pests

Accurate identification of pests is essential for effective pest management. This section will cover the identification and characteristics of common garden pests.

### Insects

### Aphids

- **Appearance:** Small, soft-bodied insects that can be green, black, red, yellow, or brown.

- **Damage:** Feed on plant sap, causing curling, yellowing, and stunted growth. Produce honeydew, which attracts ants and promotes the growth of sooty mold.

- **Signs:** Clusters of aphids on new growth, distorted leaves, presence of honeydew and ants.

## Caterpillars

- **Appearance:** Larvae of moths and butterflies, often green or brown with distinct markings.

- **Damage:** Chew holes in leaves, stems, and fruit. Can defoliate plants if infestations are severe.

- **Signs:** Ragged holes in leaves, frass (insect droppings) on leaves and stems, presence of caterpillars.

## Whiteflies

- **Appearance:** Tiny, white, moth-like insects found on the undersides of leaves.

- **Damage:** Suck sap from plants, causing yellowing and wilting. Excrete honeydew, leading to sooty mold growth.

- **Signs:** Cloud of whiteflies when plants are disturbed, yellowing leaves, presence of honeydew and sooty mold.

## Spider Mites

- **Appearance:** Tiny, spider-like arachnids that can be red, yellow, green, or brown.

- **Damage:** Suck sap from plants, causing stippling, yellowing, and leaf drop. Webbing on leaves and stems.

- **Signs:** Fine webbing, stippled or yellowed leaves, presence of mites on the undersides of leaves.

## Japanese Beetles

- **Appearance:** Metallic green and bronze beetles with distinctive white tufts along their sides.

- **Damage:** Skeletonize leaves by eating the tissue between the veins. Feed on flowers and fruit.

- **Signs:** Skeletonized leaves, clusters of beetles on plants, damaged flowers and fruit.

## Other Pests

## Slugs and Snails

- **Appearance:** Soft-bodied, slimy mollusks with or without shells.

- **Damage:** Chew irregular holes in leaves, stems, and fruit. Prefer moist, shady areas.

- **Signs:** Irregular holes, slime trails on plants and soil, presence of slugs or snails.

## Rodents

- **Appearance:** Includes mice, voles, and rats, which are small, furry mammals.

- **Damage:** Gnaw on stems, roots, and fruits. Can dig tunnels and create burrows.

- **Signs:** Gnawed plant material, burrows or tunnels, presence of rodent droppings.

## Birds

- **Appearance:** Various species of birds that may peck at fruits and vegetables.

- **Damage:** Peck holes in fruits, eat seeds and seedlings.

- **Signs:** Damaged fruit, seedlings pulled from the soil, presence of birds in the garden.

# Organic Pest Control Methods

Organic pest control methods focus on using natural and environmentally friendly techniques to manage pest populations. These methods minimize the use of synthetic chemicals and promote ecological balance.

## Cultural Controls

### Crop Rotation

Rotate crops to prevent the buildup of pests and diseases associated with specific plants.

- **Benefits:** Reduces the risk of soil-borne diseases and disrupts pest life cycles.

- **Example:** Rotate tomatoes with legumes or brassicas to avoid continuous planting of nightshades.

### Intercropping

Grow different crops together to confuse pests and create a more diverse ecosystem.

- **Benefits:** Reduces pest pressure and increases biodiversity.

- **Example:** Plant carrots and onions together to repel carrot flies and onion flies.

### Resistant Varieties

Choose plant varieties that are resistant to specific pests and diseases.

- **Benefits:** Reduces the need for chemical controls and increases plant resilience.

- **Example:** Select blight-resistant tomato varieties to prevent fungal infections.

## Biological Controls

### Beneficial Insects

Introduce or enhance populations of beneficial insects that prey on pests.

- **Ladybugs:** Feed on aphids, mites, and other small insects.

- **Praying Mantises:** Predators of a wide range of pests, including caterpillars and beetles.

- **Parasitic Wasps:** Lay eggs inside or on pests, killing them as the wasp larvae develop.

### Microbial Insecticides

Use naturally occurring microorganisms to control pest populations.

- **Bacillus thuringiensis (Bt):** A bacterium that produces toxins lethal to caterpillars and other insect larvae.

- **Beauveria bassiana:** A fungus that infects and kills various insect pests.

## Physical Controls

### Barriers

Use physical barriers to prevent pests from reaching plants.

- **Row Covers:** Lightweight fabric that covers plants, protecting them from insects and birds.

- **Netting:** Mesh material that excludes larger pests like birds and rabbits.

## Traps

Use traps to capture and monitor pest populations.

- **Sticky Traps:** Adhesive surfaces that capture flying insects like aphids and whiteflies.

- **Pheromone Traps:** Attract and trap specific pests using pheromones.

## Manual Removal

Physically remove pests from plants to reduce their populations.

- **Handpicking:** Remove pests like caterpillars, beetles, and slugs by hand.

- **Water Spray:** Use a strong jet of water to dislodge pests like aphids from plants.

## Chemical Controls

## Botanical Insecticides

Use insecticides derived from plants to control pests.

- **Neem Oil:** Derived from the neem tree, neem oil disrupts insect growth and feeding.

- **Pyrethrin:** Extracted from chrysanthemum flowers, pyrethrin targets a wide range of insects.

## Horticultural Oils

Use refined oils to smother and kill pests.

- **Dormant Oil:** Applied during the dormant season to control overwintering pests.

- **Summer Oil:** Applied during the growing season to control active pests like mites and aphids.

## Insecticidal Soaps

Use soaps to disrupt the cell membranes of soft-bodied insects.

- **Potassium Salts of Fatty Acids:** Effective against aphids, mites, and whiteflies.

## Organic Control Techniques for Common Pests

## Aphids

- **Beneficial Insects:** Release ladybugs or lacewings to prey on aphids.

- **Neem Oil:** Spray neem oil to disrupt aphid feeding and

reproduction.

- **Companion Planting:** Plant garlic, chives, or nasturtiums to repel aphids.

## Caterpillars

- **Bacillus thuringiensis (Bt):** Apply Bt to control caterpillars on vegetables and fruits.

- **Handpicking:** Remove caterpillars by hand and destroy them.

- **Row Covers:** Use row covers to prevent moths from laying eggs on plants.

## Whiteflies

- **Sticky Traps:** Place yellow sticky traps near plants to capture adult whiteflies.

- **Neem Oil:** Spray neem oil to reduce whitefly populations.

- **Companion Planting:** Plant marigolds or nasturtiums to deter whiteflies.

## Spider Mites

- **Beneficial Insects:** Introduce predatory mites to control spider mite populations.

- **Insecticidal Soap:** Spray insecticidal soap to kill mites on contact.

- **Humidity:** Increase humidity around plants to create unfavorable conditions for spider mites.

## Japanese Beetles

- **Handpicking:** Collect beetles by hand and drop them into a bucket of soapy water.

- **Neem Oil:** Spray neem oil to reduce beetle feeding and reproduction.

- **Traps:** Use pheromone traps to capture adult beetles.

## Slugs and Snails

- **Handpicking:** Remove slugs and snails by hand, especially in the evening or after rain.

- **Barriers:** Place copper tape or diatomaceous earth around plants to deter slugs and snails.

- **Beer Traps:** Set shallow dishes filled with beer to attract and drown slugs.

# Beneficial Insects and Natural Predators

Beneficial insects and natural predators play a crucial role in maintaining ecological balance and controlling pest populations. Encouraging these allies in your garden can reduce the need for chemical interventions and promote healthy plant growth.

## Beneficial Insects

### Ladybugs (Ladybird Beetles)

- **Diet:** Aphids, mites, and other small insects.

- **Life Cycle:** Adults and larvae are voracious predators, consuming large numbers of pests.

- **Attraction:** Plant flowers like dill, fennel, and yarrow to attract ladybugs.

### Lacewings

- **Diet:** Aphids, caterpillars, and other soft-bodied insects.

- **Life Cycle:** Larvae, known as "aphid lions," are particularly effective predators.

- **Attraction:** Plant flowers like cosmos, dill, and coreopsis to attract lacewings.

### Parasitic Wasps

- **Diet:** Lay eggs inside or on pests, with larvae consuming the host.

- **Target Pests:** Aphids, caterpillars, whiteflies, and beetles.

- **Attraction:** Plant nectar-rich flowers like alyssum, fennel, and dill to support parasitic wasps.

## Hoverflies (Syrphid Flies)

- **Diet:** Larvae feed on aphids and other small insects.

- **Life Cycle:** Adults feed on nectar and pollen, while larvae prey on pests.

- **Attraction:** Plant flowers like marigolds, sunflowers, and cosmos to attract hoverflies.

## Predatory Beetles

- **Diet:** A wide range of pests, including caterpillars, aphids, and slugs.

- **Types:** Ground beetles and rove beetles are common predatory beetles.

- **Attraction:** Provide ground cover and mulch to create habitats for predatory beetles.

## Encouraging Beneficial Insects

Creating a garden environment that attracts and supports beneficial insects involves several strategies.

## Plant Diversity

Plant a diverse range of flowers, herbs, and vegetables to provide habitat and food sources for beneficial insects.

- **Nectar and Pollen:** Plant a variety of flowers that produce nectar and pollen throughout the growing season.

- **Habitat:** Include plants that provide shelter and overwintering sites for beneficial insects.

## Avoiding Harmful Pesticides

Minimize the use of broad-spectrum pesticides that can harm beneficial insects along with pests.

- **Selective Pesticides:** Use targeted organic pesticides that are less harmful to beneficial insects.

- **Integrated Approach:** Combine physical, cultural, and biological controls to reduce reliance on chemical interventions.

## Providing Shelter

Create habitats that offer shelter and breeding sites for beneficial insects.

- **Mulch:** Use mulch to provide cover for ground-dwelling predators like beetles.

- **Insect Hotels:** Install insect hotels to provide nesting sites for solitary bees and predatory insects.

## Natural Predators

Natural predators, such as birds, bats, and amphibians, can also help control pest populations in the garden.

## Birds

Many bird species feed on insects, making them valuable allies in pest control.

- **Attracting Birds:** Provide bird feeders, bird baths, and nesting sites to attract birds to your garden.

- **Beneficial Species:** Encourage insectivorous birds like wrens, chickadees, and bluebirds.

## Bats

Bats are effective predators of night-flying insects, including moths and beetles.

- **Attracting Bats:** Install bat houses to provide roosting sites for bats.

- **Beneficial Species:** Encourage insect-eating bats like the little brown bat and the big brown bat.

## Amphibians

Frogs and toads feed on a variety of garden pests, including insects and slugs.

- **Attracting Amphibians:** Create water features like ponds or small water gardens to provide habitats for amphibians.

- **Beneficial Species:** Encourage native species like the American toad and green frog.

## Creating a Balanced Ecosystem

A healthy and balanced garden ecosystem supports beneficial insects and natural predators, reducing the need for chemical interventions.

## Plant Selection

Choose a mix of plants that support a diverse range of beneficial organisms.

- **Native Plants:** Incorporate native plants that are well-adapted to your local environment and support local wildlife.

- **Flowering Plants:** Plant flowers that bloom at different times to provide a continuous food source for pollinators and beneficial insects.

## Soil Health

Healthy soil promotes strong plant growth and supports a diverse soil ecosystem.

- **Organic Matter:** Add compost and other organic matter to improve soil fertility and structure.

- **Mulching:** Use mulch to conserve moisture, suppress weeds, and provide habitat for ground-dwelling predators.

## Water Management

Proper water management supports plant health and provides habitats for beneficial organisms.

- **Irrigation:** Use drip irrigation or soaker hoses to provide consistent moisture without waterlogging the soil.

- **Water Features:** Create small water features to attract amphibians and provide drinking water for birds.

## Monitoring and Adaptation

Regularly monitor your garden for pest and beneficial insect activity, and adapt your management strategies as needed.

- **Observation:** Keep track of pest populations, plant health, and the presence of beneficial insects.

- **Flexibility:** Be prepared to adjust your IPM plan based on observations and changing conditions.

Effective pest and disease management is crucial for maintaining a healthy and productive garden. Integrated Pest Management (IPM) provides a comprehensive approach that combines prevention, monitoring, identification, threshold setting, and a variety of control methods to manage pests sustainably. Identifying common pests and understanding their life cycles is essential for selecting appropriate control measures. Organic pest control methods, including cultural, biological, physical, and chemical controls, offer environmentally friendly alternatives to synthetic pesticides. Beneficial insects and natural predators play a vital role in managing pest populations and maintaining ecological balance.

# Chapter 9

# Disease Prevention and Management

Preventing and managing plant diseases is crucial for maintaining a healthy and productive garden. This section will cover the importance of disease prevention, the identification and management of common plant diseases, organic disease control methods, and strategies for maintaining overall plant health.

## Common Plant Diseases

Understanding common plant diseases and their symptoms is the first step in effective disease management. Plant diseases can be caused by various pathogens, including fungi, bacteria, viruses, and nematodes.

### Fungal Diseases

### Powdery Mildew

- **Symptoms:** White or gray powdery growth on leaves, stems, and buds. Infected leaves may become distorted and fall off prematurely.

- **Conditions:** Thrives in dry conditions with high humidity. Spreads rapidly in warm, dry weather.

- **Affected Plants:** Common in vegetables (squash, cucumbers, tomatoes), fruits (grapes, apples), and ornamentals (roses, zinnias).

## Downy Mildew

- **Symptoms:** Yellow or pale green spots on the upper leaf surface, with a downy white, gray, or purple growth on the underside. Leaves may wilt and die.

- **Conditions:** Prefers cool, moist conditions. Spreads in wet weather.

- **Affected Plants:** Affects a wide range of plants, including vegetables (lettuce, spinach), fruits (grapes), and ornamentals (impatiens).

## Rust

- **Symptoms:** Small, orange, yellow, or brown pustules on the undersides of leaves. Infected leaves may yellow and drop prematurely.

- **Conditions:** Thrives in moist conditions. Spread by wind and rain.

- **Affected Plants:** Common in beans, corn, roses, and daylilies.

## Botrytis (Gray Mold)

- **Symptoms:** Gray, fuzzy mold on flowers, leaves, stems, and fruit. Infected plant parts may become brown and mushy.

- **Conditions:** Prefers cool, damp conditions. Spreads rapidly in wet weather.

- **Affected Plants:** Affects a wide range of plants, including vegetables (tomatoes, beans), fruits (strawberries, grapes), and ornamentals (peonies, geraniums).

## Anthracnose

- **Symptoms:** Dark, sunken lesions on leaves, stems, flowers, and fruit. Infected areas may develop a pink or orange spore mass.

- **Conditions:** Thrives in warm, moist conditions. Spread by water splashing.

- **Affected Plants:** Common in beans, cucumbers, tomatoes, and ornamentals (dogwood, sycamore).

## Bacterial Diseases

## Bacterial Leaf Spot

- **Symptoms:** Water-soaked spots on leaves that turn brown or black. Spots may have a yellow halo. Leaves may drop prematurely.

- **Conditions:** Thrives in warm, wet conditions. Spread by water splashing and contaminated tools.

- **Affected Plants:** Common in peppers, tomatoes, beans, and ornamentals (geraniums, impatiens).

## Bacterial Wilt

- **Symptoms:** Sudden wilting and yellowing of leaves, often starting on one side of the plant. Stems may ooze a milky, bacterial slime when cut.

- **Conditions:** Spread by insects (e.g., cucumber beetles) and infected soil.

- **Affected Plants:** Common in cucumbers, melons, and squash.

## Fire Blight

- **Symptoms:** Blackened, shriveled blossoms, leaves, and twigs, giving a "burned" appearance. Affected areas may ooze a bacterial slime.

- **Conditions:** Thrives in warm, humid conditions. Spread by insects, rain, and pruning tools.

- **Affected Plants:** Common in apples, pears, and related ornamentals (hawthorn, crabapple).

## Viral Diseases

## Tomato Mosaic Virus (ToMV)

- **Symptoms:** Mottled, light and dark green patterns on leaves. Leaves may become distorted and stunted. Fruit may develop yellow streaks or rings.

- **Conditions:** Spread by infected seeds, plants, and tools.

- **Affected Plants:** Common in tomatoes, peppers, and some ornamentals (petunias).

## Cucumber Mosaic Virus (CMV)

- **Symptoms:** Mottled yellow and green leaves, with leaf distortion and stunted growth. Fruit may develop warty or mottled appearance.

- **Conditions:** Spread by aphids and infected plants.

- **Affected Plants:** Common in cucumbers, melons, squash, and ornamentals (lilies, dahlias).

## Nematode Diseases

## Root-Knot Nematodes

- **Symptoms:** Knobby, swollen roots, stunted growth, yellowing, and wilting. Plants may be more susceptible to other diseases.

- **Conditions:** Thrives in warm, sandy soils. Spread by infected soil and plants.

- **Affected Plants:** Common in tomatoes, carrots, beans, and many other vegetables.

## Organic Disease Control Methods

Organic disease control methods focus on preventing diseases through cultural practices, biological controls, and organic treatments. These methods aim to create a healthy garden environment that minimizes disease development.

## Cultural Controls

### Crop Rotation

Rotate crops to prevent the buildup of disease pathogens in the soil.

- **Benefits:** Reduces the risk of soil-borne diseases and disrupts disease life cycles.

- **Example:** Rotate tomatoes with legumes or brassicas to avoid continuous planting of nightshades.

### Sanitation

Maintain a clean garden to reduce the spread of disease pathogens.

- **Remove Debris:** Regularly remove and dispose of plant debris, fallen leaves, and diseased plants.

- **Clean Tools:** Disinfect gardening tools regularly, especially after working with diseased plants.

- **Mulching:** Use organic mulch to prevent soil from splashing onto plants and spreading disease pathogens.

## Proper Watering

Water plants properly to reduce the risk of diseases.

- **Avoid Overwatering:** Water plants deeply but infrequently to promote strong root growth and prevent waterlogging.

- **Water at the Base:** Water at the base of plants to keep foliage dry and reduce the risk of fungal diseases.

- **Drip Irrigation:** Use drip irrigation or soaker hoses to deliver water directly to the soil, minimizing leaf wetness.

## Spacing and Air Circulation

Provide adequate spacing and air circulation to reduce humidity and prevent disease.

- **Proper Spacing:** Space plants according to their needs to allow for good air circulation.

- **Pruning:** Prune plants to remove overcrowded or damaged branches and improve airflow.

## Biological Controls

## Beneficial Microorganisms

Introduce beneficial microorganisms to the soil to suppress disease pathogens.

- **Mycorrhizal Fungi:** Form symbiotic relationships with plant roots, enhancing nutrient uptake and disease resistance.

- **Trichoderma:** Fungi that compete with and inhibit disease-causing fungi in the soil.

## Disease-Resistant Varieties

Plant disease-resistant varieties to reduce susceptibility to specific diseases.

- **Example:** Choose blight-resistant tomato varieties to prevent fungal infections.

## Companion Planting

Use companion planting to reduce disease pressure and promote plant health.

- **Example:** Plant garlic or onions near susceptible plants to repel fungal pathogens.

## Organic Treatments

## Neem Oil

Neem oil is a natural fungicide and insecticide derived from the neem tree.

- **Application:** Mix neem oil with water and apply as a foliar spray to control fungal diseases like powdery mildew and rust.

- **Benefits:** Neem oil also repels and kills various insect pests.

## Baking Soda

Baking soda can be used to prevent and control fungal diseases.

- **Application:** Mix 1 tablespoon of baking soda with 1 gallon of water and a few drops of liquid soap. Apply as a foliar spray to control powdery mildew and other fungal diseases.

- **Benefits:** Baking soda raises the pH on leaf surfaces, creating an inhospitable environment for fungi.

## Copper Fungicides

Copper fungicides are effective against a range of fungal and bacterial diseases.

- **Application:** Apply copper fungicides as a foliar spray according to label instructions to control diseases like blight, anthracnose, and bacterial leaf spot.

- **Benefits:** Copper fungicides provide broad-spectrum disease control but should be used sparingly to avoid copper buildup in the soil.

## Horticultural Oils

Horticultural oils can control fungal diseases and some insect pests.

- **Application:** Apply horticultural oils as a foliar spray to control diseases like powdery mildew and rust.

- **Benefits:** Horticultural oils also smother insect pests like aphids and mites.

## Maintaining Plant Health

Maintaining overall plant health is crucial for preventing diseases and ensuring a productive garden. Healthy plants are more resilient and better able to withstand pest and disease pressures.

## Soil Health

Healthy soil supports strong plant growth and disease resistance.

## Organic Matter

Add organic matter to improve soil structure, fertility, and microbial activity.

- **Compost:** Regularly add compost to the soil to provide nutrients and improve soil health.

- **Mulch:** Use organic mulch to conserve moisture, suppress weeds, and add organic matter as it decomposes.

## Soil Testing

Conduct soil tests to monitor nutrient levels and pH.

- **Nutrient Management:** Adjust fertilization based on soil test results to provide balanced nutrition for plants.

- **pH Adjustment:** Amend the soil to maintain an optimal pH for plant growth.

# Fertilization

Provide adequate nutrition to support healthy plant growth.

## Organic Fertilizers

Use organic fertilizers to provide essential nutrients without harming soil health.

- **Compost:** Incorporate compost into the soil to provide a slow-release source of nutrients.

- **Manure:** Apply well-rotted manure to improve soil fertility and structure.

- **Organic Fertilizer Blends:** Use commercially available organic fertilizer blends to provide balanced nutrition.

## Balanced Fertilization

Avoid over-fertilization, which can lead to excessive vegetative growth and increased disease susceptibility.

- **Slow-Release Fertilizers:** Use slow-release fertilizers to provide a steady supply of nutrients over time.

- **Fertilizer Timing:** Apply fertilizers at appropriate times during the growing season to support plant growth and development.

## Irrigation

Proper irrigation practices are essential for maintaining plant health.

## Consistent Moisture

Provide consistent moisture to support healthy root development.

- **Deep Watering:** Water deeply but infrequently to encourage deep root growth.

- **Mulching:** Use mulch to retain soil moisture and reduce the need for frequent watering.

## Avoiding Water Stress

Prevent water stress, which can weaken plants and increase disease susceptibility.

- **Regular Monitoring:** Monitor soil moisture levels regularly and adjust irrigation as needed.

- **Drought-Tolerant Plants:** Choose drought-tolerant plants for areas prone to water stress.

## Pruning and Training

Proper pruning and training techniques promote healthy plant growth and reduce disease risk.

## Pruning

Prune plants to remove dead, diseased, or overcrowded branches.

- **Disease Prevention:** Remove diseased plant parts promptly to prevent the spread of pathogens.

- **Air Circulation:** Prune to improve air circulation and reduce humidity around plants.

## Training

Train plants to support healthy growth and reduce disease risk.

- **Staking and Trellising:** Support vining and tall plants with stakes or trellises to keep foliage off the ground and improve air circulation.

- **Espalier:** Train fruit trees and shrubs against a wall or fence to improve sunlight exposure and air circulation.

## Pest Management

Effective pest management reduces the risk of pest-related diseases.

### Integrated Pest Management (IPM)

Implement IPM practices to manage pest populations and reduce disease risk.

- **Monitoring:** Regularly monitor for pests and take action when populations reach threshold levels.

- **Biological Controls:** Use beneficial insects and natural predators to control pest populations.

- **Cultural Controls:** Adjust cultural practices, such as crop rotation and intercropping, to reduce pest pressure.

## Organic Pesticides

Use organic pesticides as a last resort to manage pests and reduce disease risk.

- **Neem Oil:** Use neem oil to control a wide range of insect pests and fungal diseases.

- **Insecticidal Soap:** Apply insecticidal soap to control soft-bodied insects like aphids and whiteflies.

## Environmental Management

Manage the garden environment to create conditions that support plant health and reduce disease risk.

## Sunlight

Ensure plants receive adequate sunlight to support healthy growth and reduce disease risk.

- **Site Selection:** Choose garden sites with full sun exposure for most vegetables and fruits.

- **Shade Management:** Provide shade for plants that require partial shade to prevent heat stress.

## Air Circulation

Improve air circulation to reduce humidity and prevent fungal diseases.

- **Spacing:** Space plants appropriately to allow for good air circulation.

- **Pruning:** Prune plants to remove overcrowded branches and improve airflow.

## Temperature Management

Manage temperature extremes to support plant health and reduce stress.

- **Mulching:** Use mulch to regulate soil temperature and protect plant roots.

- **Row Covers:** Use row covers or shade cloth to protect plants from extreme heat or cold.

## Advanced Disease Prevention Strategies

For experienced gardeners, advanced disease prevention strategies can further enhance plant health and productivity.

## Grafting

Grafting involves joining the tissue of one plant (the scion) to the root system of another plant (the rootstock).

- **Benefits:** Grafting can provide disease resistance, improve plant vigor, and enhance productivity.

- **Example:** Graft heirloom tomato varieties onto disease-resistant rootstocks to prevent soil-borne diseases like fusarium wilt and nematodes.

## Solarization

Soil solarization uses solar heat to kill soil-borne pathogens and pests.

- **Method:** Cover the soil with clear plastic for 4-6 weeks during the hottest part of the summer. The heat trapped under the plastic kills pathogens and pests.

- **Benefits:** Solarization is an effective organic method for reducing soil-borne diseases and improving soil health.

## Biofumigation

Biofumigation uses certain plants, like mustards and radishes, to release natural compounds that suppress soil-borne pathogens.

- **Method:** Grow biofumigant crops and then chop and incorporate them into the soil. The breakdown of plant tissues releases bioactive compounds that suppress pathogens.

- **Benefits:** Biofumigation improves soil health and reduces the incidence of soil-borne diseases.

## Companion Planting

Companion planting involves growing certain plants together to reduce disease pressure and promote plant health.

- **Benefits:** Companion plants can repel pests, attract beneficial insects, and improve soil health.

- **Example:** Plant marigolds with tomatoes to repel nematodes and improve tomato health.

Effective disease prevention and management are crucial for maintaining a healthy and productive garden. By understanding common plant diseases and their symptoms, implementing organic disease control methods, and maintaining overall plant health, gardeners can reduce disease pressure and enhance plant resilience. Integrating cultural, biological, and organic treatments into a comprehensive disease management plan will help create a thriving garden ecosystem that supports strong, healthy plants and bountiful harvests. By focusing on prevention and promoting plant health, gardeners can enjoy the benefits of a productive and sustainable garden for years to come.

# Chapter 10

# Raising Backyard Chickens

Raising backyard chickens is a rewarding project that offers many benefits, including fresh eggs, natural pest control, and the joy of watching these lively creatures. This chapter will guide you through the essentials of choosing the right chicken breeds, building a suitable chicken coop, and feeding and caring for your chickens.

## Choosing the Right Breeds

Selecting the right breed of chickens is crucial for achieving your goals, whether they are egg production, meat production, or simply enjoying the companionship of these fascinating birds. Different breeds have varying characteristics in terms of egg production, temperament, and hardiness.

## Egg-Laying Breeds

For those primarily interested in egg production, several breeds are known for their high egg-laying capabilities.

1. **Rhode Island Red:**

   - **Egg Production:** 250-300 brown eggs per year.

   - **Temperament:** Hardy, adaptable, and generally friendly.

   - **Description:** Red feathers with a robust build.

2. **Leghorn:**

   - **Egg Production:** 280-320 white eggs per year.

   - **Temperament:** Active and slightly skittish.

   - **Description:** White feathers and a sleek body.

3. **Plymouth Rock:**

   - **Egg Production:** 200-280 brown eggs per year.

   - **Temperament:** Friendly, docile, and good with children.

   - **Description:** Black and white barred feathers with a sturdy frame.

4. **Sussex:**

   - **Egg Production:** 250-280 brown eggs per year.

   - **Temperament:** Gentle and curious.

   - **Description:** White feathers with black markings on the neck and tail.

## Dual-Purpose Breeds

Dual-purpose breeds are suitable for both egg production and meat. These breeds offer a balance of productivity and hardiness.

1. **Orpington:**

   - **Egg Production:** 200-280 brown eggs per year.

   - **Meat Production:** Good for meat due to their larger size.

   - **Temperament:** Gentle, friendly, and easy to handle.

   - **Description:** Buff, black, blue, or white feathers with a large body.

2. **Australorp:**

   - **Egg Production:** 250-300 brown eggs per year.

   - **Meat Production:** Suitable for meat due to their size.

   - **Temperament:** Calm and friendly.

   - **Description:** Black feathers with a greenish sheen.

3. **Wyandotte:**

   - **Egg Production:** 200-240 brown eggs per year.

   - **Meat Production:** Good for meat due to their solid build.

   - **Temperament:** Docile and good with children.

- **Description:** Silver-laced, gold-laced, or other color varieties with a rose comb.

## Meat Breeds

For those focused on meat production, certain breeds are bred specifically for rapid growth and larger body size.

1. **Cornish Cross:**

   - **Meat Production:** Fast-growing and ready for processing in 6-8 weeks.

   - **Temperament:** Docile but less active due to rapid growth.

   - **Description:** White feathers with a broad breast and short legs.

2. **Jersey Giant:**

   - **Meat Production:** Slow-growing but reaches large sizes, ideal for meat.

   - **Temperament:** Calm and friendly.

   - **Description:** Black, white, or blue feathers with a robust build.

## Ornamental Breeds

For those who enjoy the aesthetic and companionship aspects of raising chickens, ornamental breeds offer unique appearances and charming personalities.

1. **Silkie:**

   ◦ **Egg Production:** 100-120 small cream eggs per year.

   ◦ **Temperament:** Exceptionally friendly and good with children.

   ◦ **Description:** Fluffy, silk-like feathers in various colors, with a small size and black skin.

2. **Polish:**

   ◦ **Egg Production:** 150-200 white eggs per year.

   ◦ **Temperament:** Friendly but slightly skittish.

   ◦ **Description:** Striking crest of feathers on the head, available in various colors.

3. **Cochin:**

   ◦ **Egg Production:** 150-180 brown eggs per year.

   ◦ **Temperament:** Gentle and docile.

   ◦ **Description:** Large size with feathered legs and a variety of colors.

## Building a Chicken Coop

A well-designed chicken coop is essential for the health and safety of your chickens. The coop should provide shelter from the elements, protection from predators, and a comfortable environment for laying eggs and roosting.

## Planning the Coop

Before building the coop, consider the following factors:

1. **Number of Chickens:**

   ○ **Space Requirements:** Provide at least 4 square feet of indoor space per chicken and 8-10 square feet of outdoor run space per chicken.

2. **Climate:**

   ○ **Ventilation:** Ensure adequate ventilation to prevent moisture buildup and ammonia accumulation.

   ○ **Insulation:** In colder climates, insulate the coop to keep chickens warm during winter.

3. **Location:**

   ○ **Sunlight:** Place the coop in a location that receives morning sunlight and some afternoon shade.

   ○ **Drainage:** Choose a well-drained area to prevent water accumulation around the coop.

## Materials and Tools

Gather the necessary materials and tools for building the coop.

1. **Materials:**

   ○ **Wood:** Pressure-treated lumber for the frame, plywood for the walls and floor, and hardware cloth for windows and vents.

- ○ **Roofing:** Metal or asphalt shingles for the roof.

- ○ **Nesting Boxes:** Wood or plastic for constructing nesting boxes.

- ○ **Roosts:** Wooden dowels or branches for roosting bars.

- ○ **Hardware:** Hinges, latches, screws, nails, and door handles.

2. **Tools:**

- ○ **Measuring Tape:** For accurate measurements.

- ○ **Saw:** Circular saw or hand saw for cutting wood.

- ○ **Drill:** For making holes and driving screws.

- ○ **Hammer:** For driving nails.

- ○ **Level:** For ensuring the coop is level.

- ○ **Screwdriver:** For fastening screws.

## Building the Coop

Follow these steps to build a functional and safe chicken coop.

1. **Foundation:**

- ○ **Level the Ground:** Clear and level the ground where the coop will be built.

- ○ **Foundation Options:** Use concrete blocks, pavers, or treated wood skids to create a sturdy foundation.

2. **Frame:**

- **Construct the Base:** Build a rectangular base using pressure-treated lumber.

- **Build the Walls:** Construct the walls by attaching vertical studs to the base frame. Include openings for windows and doors.

- **Roof Structure:** Add rafters to support the roof. Ensure the roof has a slight pitch for water drainage.

3. **Walls and Roof:**

- **Attach Plywood:** Cover the walls and roof with plywood, securing it with screws or nails.

- **Install Roofing Material:** Attach metal sheets or asphalt shingles to the roof.

4. **Ventilation:**

- **Windows and Vents:** Cut openings for windows and vents. Cover them with hardware cloth to keep predators out.

- **Vent Placement:** Place vents near the roofline to allow hot air to escape and prevent drafts at chicken level.

5. **Nesting Boxes and Roosts:**

- **Nesting Boxes:** Build and install nesting boxes inside the coop. Provide one nesting box for every 3-4 hens.

- **Roosts:** Install roosting bars at varying heights, ensuring they are higher than the nesting boxes to encourage roosting.

6. **Doors and Security:**

- **Chicken Door:** Install a small door for chickens to access the outdoor run.

- **Human Door:** Include a larger door for easy access to the coop for cleaning and maintenance.

- **Predator Proofing:** Ensure all openings are secure and install latches on doors to keep predators out.

7. **Outdoor Run:**

- **Enclosure:** Build an outdoor run with sturdy fencing, such as hardware cloth or welded wire, to provide a safe area for chickens to roam.

- **Shade and Shelter:** Provide shaded areas and shelter within the run to protect chickens from the sun and rain.

## Feeding and Caring for Chickens

Proper feeding and care are essential for the health and productivity of your chickens. This includes providing a balanced diet, clean water, regular health checks, and maintaining a clean environment.

## Feeding Chickens

Chickens require a balanced diet to produce healthy eggs and maintain their overall health.

## Commercial Feeds

Commercial chicken feeds are formulated to provide the necessary nutrients for different stages of a chicken's life.

1. **Starter Feed:**

   - **Age:** From hatch to 6 weeks old.

   - **Protein:** High protein content (18-20%) to support rapid growth.

   - **Form:** Crumbles or mash for easy consumption.

2. **Grower Feed:**

   - **Age:** From 6 to 20 weeks old.

   - **Protein:** Moderate protein content (16-18%) to support continued growth.

   - **Form:** Crumbles or pellets.

3. **Layer Feed:**

   - **Age:** From 20 weeks old and onward (or when hens start laying).

   - **Protein:** Lower protein content (16-18%) with added calcium for eggshell formation.

- **Form:** Pellets or crumbles.

## 4. Broiler Feed:

- **Age:** For meat chickens.

- **Protein:** High protein content (20-24%) to support rapid growth.

- **Form:** Pellets or crumbles.

## Supplemental Feeds

In addition to commercial feed, chickens can benefit from various supplemental feeds.

### 1. Scratch Grains:

- **Purpose:** Treat and foraging stimulation.

- **Content:** Mixture of cracked corn, wheat, barley, and oats.

- **Feeding:** Offer in moderation to prevent obesity.

### 2. Kitchen Scraps:

- **Acceptable Scraps:** Vegetables, fruits, grains, and cooked pasta.

- **Avoid:** Raw potatoes, avocado, chocolate, and high-salt or high-fat foods.

### 3. Grit:

- **Purpose:** Aid in digestion by grinding food in the gizzard.

- **Types:** Insoluble grit (granite) and soluble grit (oyster shell).

- **Feeding:** Provide in a separate container free-choice.

4. **Calcium Supplements:**

- **Purpose:** Ensure strong eggshells.

- **Sources:** Crushed oyster shells or commercial calcium supplements.

- **Feeding:** Provide in a separate container free-choice.

## Feeding Tips

1. **Clean Feeders:** Clean feeders regularly to prevent mold and contamination.

2. **Fresh Feed:** Store feed in a cool, dry place to maintain freshness and prevent spoilage.

3. **Free-Range Foraging:** Allow chickens to forage for insects, greens, and seeds to supplement their diet.

## Providing Clean Water

Access to clean, fresh water is essential for chicken health and productivity.

1. **Water Containers:**

- **Types:** Waterers or automatic water systems.

- **Placement:** Keep water containers elevated to prevent

contamination.

## 2. Cleanliness:

- ○ **Daily Cleaning:** Clean water containers daily to prevent algae and bacteria growth.

- ○ **Weekly Scrubbing:** Scrub water containers with a mild bleach solution weekly to disinfect.

## 3. Fresh Water:

- ○ **Refilling:** Refill water containers regularly to ensure a continuous supply of fresh water.

## Health Care and Maintenance

Regular health care and maintenance are crucial for preventing diseases and ensuring the well-being of your chickens.

## Health Checks

Perform regular health checks to monitor your chickens' condition and catch any issues early.

1. **Appearance:** Check for bright eyes, clean feathers, and alert behavior.

2. **Weight:** Monitor weight to ensure chickens are maintaining a healthy body condition.

3. **Feet and Legs:** Inspect feet and legs for cuts, swelling, or infections.

4. **Comb and Wattles:** Look for color and texture changes that might indicate illness.

## Parasite Control

Control parasites to prevent health problems and ensure the well-being of your flock.

1. **External Parasites:**

   ○ **Lice and Mites:** Regularly inspect chickens for lice and mites. Treat infestations with poultry dust or sprays.

   ○ **Dust Baths:** Provide a dust bath area with sand, diatomaceous earth, or wood ash to help chickens keep external parasites at bay.

2. **Internal Parasites:**

   ○ **Worms:** Monitor for signs of internal parasites, such as weight loss and diarrhea. Use deworming medications as needed.

## Vaccinations

Consider vaccinating your chickens to protect them from common diseases.

1. **Marek's Disease:** A common viral disease in chickens. Vaccinate chicks at the hatchery or within the first few days of life.

2. **Other Vaccinations:** Consult with a veterinarian or poultry specialist about other recommended vaccinations based on local disease prevalence.

## Coop Maintenance

Regular coop maintenance is essential for a healthy environment.

1. **Cleaning:**

   ○ **Daily Tasks:** Remove droppings from nesting boxes and roosting areas.

   ○ **Weekly Tasks:** Replace bedding, clean feeders and waterers, and inspect the coop for damage or wear.

   ○ **Monthly Tasks:** Deep clean the coop by removing all bedding, scrubbing surfaces, and disinfecting with a mild bleach solution.

2. **Bedding:**

   ○ **Types:** Use straw, wood shavings, or other absorbent materials for bedding.

   ○ **Replacement:** Change bedding regularly to maintain a clean, dry environment.

## Protecting from Predators

Ensure the coop and run are secure to protect chickens from predators.

1. **Predator-Proofing:**

   ○ **Fencing:** Use sturdy, buried fencing to prevent digging predators.

   ○ **Locks and Latches:** Use secure locks on coop doors and

windows.

- **Covering:** Cover the top of the run with wire or netting to prevent aerial predators.

## Seasonal Care

Adapting care routines to seasonal changes ensures chickens remain healthy and productive throughout the year.

### Winter Care

1. **Cold Protection:**

   - **Insulation:** Insulate the coop to maintain a warm environment.

   - **Heat Sources:** Use heat lamps or heaters if necessary, but ensure they are safe and well-secured.

2. **Water:**

   - **Prevent Freezing:** Use heated water containers or change water frequently to prevent freezing.

3. **Diet:**

   - **Increased Calories:** Provide additional calories with high-energy feeds like cracked corn to help chickens maintain body heat.

### Summer Care

1. **Heat Protection:**

   ○ **Shade:** Provide shaded areas in the run to protect chickens from direct sunlight.

   ○ **Ventilation:** Ensure adequate ventilation in the coop to prevent overheating.

2. **Water:**

   ○ **Cool Water:** Provide fresh, cool water throughout the day to prevent dehydration.

3. **Diet:**

   ○ **Electrolytes:** Add electrolytes to water during extreme heat to help chickens stay hydrated.

## Brooding and Raising Chicks

If you plan to hatch and raise chicks, proper brooding and care are essential for their survival and growth.

## Setting Up the Brooder

1. **Brooder Box:**

   ○ **Size:** Provide at least 0.5 square feet per chick for the first few weeks.

   ○ **Material:** Use a cardboard box, plastic tub, or wooden crate.

2. **Heat Source:**

- **Heat Lamp:** Use a heat lamp to maintain a temperature of 95°F for the first week, reducing by 5°F each week until fully feathered.

- **Placement:** Ensure the heat lamp is securely positioned to avoid fire hazards.

3. **Bedding:**

- **Material:** Use pine shavings or straw for bedding.

- **Maintenance:** Change bedding regularly to keep the brooder clean and dry.

4. **Feed and Water:**

- **Feed:** Provide chick starter feed with 18-20% protein.

- **Water:** Provide clean, fresh water in shallow containers to prevent drowning.

## Caring for Chicks

1. **Temperature Monitoring:**

- **Thermometer:** Use a thermometer to monitor brooder temperature.

- **Chick Behavior:** Observe chick behavior; if they huddle under the heat source, they are cold. If they avoid it, they are too hot.

2. **Socialization:**

- **Handling:** Gently handle chicks daily to acclimate them to human interaction.

- **Observation:** Monitor chicks for signs of illness or distress.

### 3. Growth and Development:

- **Feathering:** As chicks grow and feather out, adjust the brooder temperature accordingly.

- **Space:** Increase space as chicks grow to prevent overcrowding.

### 4. Transition to Coop:

- **Timing:** Move chicks to the coop once they are fully feathered and can regulate their body temperature (usually 6-8 weeks old).

- **Integration:** Introduce chicks to the adult flock gradually to prevent aggression.

Raising backyard chickens can be a fulfilling and productive venture, offering fresh eggs, natural pest control, and the enjoyment of caring for these lively birds. By choosing the right breeds, building a well-designed chicken coop, and providing proper feeding and care, you can ensure the health and happiness of your chickens. Additionally, understanding the importance of seasonal care and brooding chicks will help you manage your flock effectively throughout the year. With attention to detail and dedication, you can create a thriving backyard flock that enhances your homesteading experience.

# Chapter 11

# Water and Irrigation

Efficient water management is crucial for maintaining a healthy and productive homestead. Whether you're growing vegetables, fruits, or flowers, providing the right amount of water at the right time can significantly enhance plant growth and yield. This chapter will cover efficient watering techniques, the design and implementation of drip irrigation systems, and the benefits and methods of rainwater harvesting.

## Efficient Watering Techniques

Efficient watering techniques ensure that plants receive the necessary moisture without wasting water. Proper watering practices can also prevent water-related problems such as soil erosion, nutrient leaching, and plant diseases.

## Understanding Plant Water Needs

Different plants have varying water requirements based on their species, growth stage, and environmental conditions. Understanding these needs is the first step towards efficient watering.

1. **Water Requirements by Plant Type:**

   ○ **Vegetables:** Generally require consistent moisture, especially during flowering and fruiting stages. Leafy greens need more frequent watering compared to root vegetables.

   ○ **Fruits:** Fruit-bearing plants and trees need deep watering, especially during flowering and fruit development.

   ○ **Flowers:** Annuals typically require more frequent watering than perennials. Drought-tolerant species need less water once established.

   ○ **Herbs:** Most herbs prefer well-drained soil and can tolerate some dryness, but regular watering helps maintain vigorous growth.

2. **Growth Stage Considerations:**

   ○ **Seedlings:** Require frequent, light watering to keep the soil consistently moist.

   ○ **Mature Plants:** Benefit from deeper, less frequent watering to encourage deep root growth.

   ○ **Fruiting/Flowering Plants:** Need more water during peak production periods.

3. **Environmental Factors:**

   ○ **Temperature:** Higher temperatures increase evaporation and plant water needs.

   ○ **Humidity:** Lower humidity levels increase water loss from soil and plants.

- **Wind:** Windy conditions can dry out plants and soil faster, increasing water requirements.

- **Soil Type:** Sandy soils drain quickly and need more frequent watering, while clay soils retain moisture longer and require less frequent watering.

## Watering Methods

There are several methods to water your garden efficiently. Each has its advantages and is suitable for different types of plants and garden layouts.

## Hand Watering

Hand watering is suitable for small gardens, containers, and specific plants that need individual attention.

1. **Advantages:**

   - Precise control over water application.

   - Ability to target specific plants and avoid wetting foliage.

   - Immediate response to plant needs.

2. **Disadvantages:**

   - Time-consuming for large gardens.

   - Requires regular monitoring and manual labor.

### 3. **Techniques:**

- Use a watering can or a hose with a spray nozzle.

- Water the base of plants to avoid wetting foliage and reduce disease risk.

- Apply water slowly to allow it to soak into the soil and reach the root zone.

## Overhead Sprinklers

Overhead sprinklers are suitable for larger gardens, lawns, and areas with uniform plant types.

### 1. **Advantages:**

- Covers large areas quickly and evenly.

- Easy to set up and use.

### 2. **Disadvantages:**

- Water can evaporate before reaching the soil, especially in hot or windy conditions.

- Wet foliage increases the risk of fungal diseases.

- Less efficient in delivering water directly to the root zone.

### 3. **Techniques:**

- Choose the right type of sprinkler for your garden layout

(oscillating, rotary, or stationary).

- Water early in the morning or late in the evening to minimize evaporation and fungal diseases.

- Adjust the sprinkler to avoid watering paths, driveways, and other non-plant areas.

## Soaker Hoses

Soaker hoses are porous hoses that release water slowly along their length, making them ideal for row crops, flower beds, and vegetable gardens.

1. **Advantages:**

   - Delivers water directly to the soil and root zone.

   - Reduces evaporation and water waste.

   - Minimizes wetting of foliage, reducing disease risk.

2. **Disadvantages:**

   - Requires careful placement to ensure even water distribution.

   - Can be prone to clogging if not properly maintained.

3. **Techniques:**

   - Lay the soaker hose on the soil surface or cover it with mulch to reduce evaporation.

   - Connect multiple hoses with connectors to cover larger areas.

○ Use a timer to automate watering and ensure consistent moisture levels.

## Drip Irrigation

Drip irrigation systems deliver water directly to the base of plants through a network of tubes and emitters. This method is highly efficient and customizable.

1. **Advantages:**

   ○ Provides precise control over water application.

   ○ Reduces water waste through evaporation and runoff.

   ○ Minimizes wetting of foliage, reducing disease risk.

   ○ Can be automated for consistent watering.

2. **Disadvantages:**

   ○ Higher initial setup cost and complexity.

   ○ Requires regular maintenance to prevent clogging.

3. **Techniques:**

   ○ Design the system to match the layout and needs of your garden.

   ○ Use emitters with adjustable flow rates for different plant types.

   ○ Regularly check for clogs and leaks to ensure efficient operation.

## Best Practices for Efficient Watering

Adopting best practices for efficient watering can help conserve water and promote healthy plant growth.

1. **Mulching:**

   ○ Apply organic mulch (straw, wood chips, leaves) around plants to retain soil moisture, reduce evaporation, and suppress weeds.

2. **Watering Schedule:**

   ○ Water deeply but infrequently to encourage deep root growth.

   ○ Adjust watering frequency based on weather conditions and plant needs.

3. **Watering Time:**

   ○ Water early in the morning or late in the evening to minimize evaporation and reduce disease risk.

4. **Soil Moisture Monitoring:**

   ○ Use a soil moisture meter or simply check soil moisture by hand to determine when watering is needed.

5. **Group Plants by Water Needs:**

   ○ Plant species with similar water requirements together to optimize watering efficiency.

## Drip Irrigation Systems

Drip irrigation systems are among the most efficient and effective methods for watering gardens. They deliver water directly to the root zone of plants, minimizing evaporation and runoff. This section covers the design, installation, and maintenance of drip irrigation systems.

## Designing a Drip Irrigation System

Proper design is crucial for creating an efficient and effective drip irrigation system. Consider the following factors when planning your system.

1. **Water Source:**

   - Identify your water source (tap, well, or rainwater collection system) and ensure it can provide sufficient pressure and flow rate for the system.

2. **Garden Layout:**

   - Sketch a map of your garden, including plant locations, rows, beds, and paths.

   - Group plants with similar water needs together for efficient watering.

3. **System Components:**

   - **Mainline Tubing:** The primary tubing that carries water from the source to the distribution lines.

   - **Distribution Tubing:** Smaller tubing that branches off from the mainline to deliver water to specific areas.

- **Emitters:** Devices that control the flow rate and deliver water directly to the plants.

- **Connectors:** Fittings that connect different sections of tubing and emitters.

- **Filters:** Prevent debris from clogging the system.

- **Pressure Regulators:** Ensure consistent water pressure throughout the system.

- **Timers:** Automate watering schedules for consistent moisture levels.

## Installing a Drip Irrigation System

Follow these steps to install a drip irrigation system in your garden.

1. **Prepare the Site:**

   - Clear the area of weeds, debris, and obstacles.

   - Lay out the mainline tubing along the planned path.

2. **Connect the Mainline Tubing:**

   - Attach the mainline tubing to the water source using appropriate connectors and fittings.

   - Install a filter and pressure regulator at the water source to ensure clean water and consistent pressure.

3. **Lay the Distribution Tubing:**

- Run the distribution tubing from the mainline to the specific planting areas.

- Use connectors and stakes to secure the tubing in place.

### 4. Install Emitters:

- Punch holes in the distribution tubing at the desired locations and insert emitters.

- Use emitters with adjustable flow rates for different plant types.

### 5. Test the System:

- Turn on the water and check for leaks, clogs, and even water distribution.

- Adjust emitters and connectors as needed to ensure efficient operation.

### 6. Cover the Tubing:

- Cover the tubing with mulch to protect it from UV damage and reduce evaporation.

## Maintaining a Drip Irrigation System

Regular maintenance is essential for keeping your drip irrigation system functioning efficiently.

### 1. Inspect Regularly:

- Check the system for leaks, clogs, and damage.

○ Replace damaged tubing and fittings as needed.

## 2. **Clean Filters:**

○ Clean or replace filters regularly to prevent clogs and ensure clean water flow.

## 3. **Flush the System:**

○ Periodically flush the system to remove any debris or sediment that may have accumulated.

## 4. **Winterizing:**

○ In cold climates, drain the system and store components indoors to prevent damage from freezing temperatures.

## 5. **Adjust Emitters:**

○ Adjust emitters as plants grow and their water needs change.

# Rainwater Harvesting

Rainwater harvesting involves collecting and storing rainwater for use in your garden. This sustainable practice reduces reliance on municipal water supplies and provides high-quality water for plants.

## Benefits of Rainwater Harvesting

### 1. **Water Conservation:**

○ Reduces demand on municipal water supplies and conserves

water resources.

## 2. Cost Savings:

○ Lowers water bills by providing a free source of water for your garden.

## 3. Improved Plant Health:

○ Rainwater is naturally soft and free of chemicals, making it ideal for plants.

## 4. Environmental Benefits:

○ Reduces stormwater runoff, which can carry pollutants into waterways.

○ Promotes sustainability and self-sufficiency.

## Components of a Rainwater Harvesting System

A basic rainwater harvesting system consists of several key components.

### 1. Catchment Area:

○ Typically, the roof of a house, shed, or greenhouse.

○ The catchment area collects rainwater and directs it to the collection system.

### 2. Gutters and Downspouts:

○ Gutters collect rainwater from the roof and channel it to the downspouts.

- Downspouts direct water to the storage tank or collection area.

## 3. **First Flush Diverter:**

- A device that diverts the initial flow of rainwater away from the storage tank to remove debris and contaminants.

- Ensures that cleaner water enters the storage tank.

## 4. **Storage Tank:**

- A tank or barrel that stores collected rainwater.

- Tanks can be made from various materials, including plastic, metal, or concrete.

- Choose a tank size based on your water needs and available space.

## 5. **Filter:**

- A filter removes debris and contaminants from the rainwater before it enters the storage tank.

- Can be a simple mesh screen or a more complex filtration system.

## 6. **Overflow System:**

- An overflow pipe or outlet directs excess water away from the storage tank when it reaches capacity.

- Prevents flooding and damage to the tank.

7. **Delivery System:**

- A pump or gravity-fed system delivers stored rainwater to the garden.

- Can be connected to a drip irrigation system, soaker hoses, or hand watering tools.

## Setting Up a Rainwater Harvesting System

Follow these steps to set up a rainwater harvesting system for your garden.

1. **Evaluate Your Catchment Area:**

- Measure the size of your roof to estimate the potential rainwater collection.

- Clean gutters and downspouts to ensure they are free of debris.

2. **Choose a Storage Tank:**

- Select a tank size based on your water needs and available space.

- Ensure the tank is made of food-grade material if using collected water for edible plants.

3. **Install Gutters and Downspouts:**

- Install or repair gutters and downspouts to direct rainwater from the roof to the storage tank.

- Use gutter guards to prevent leaves and debris from entering the system.

### 4. Install a First Flush Diverter:

- Attach the first flush diverter to the downspout to divert the initial flow of rainwater.

### 5. Set Up the Storage Tank:

- Place the storage tank on a stable, level surface near the downspout.

- Connect the downspout to the tank inlet.

- Install a filter at the tank inlet to remove debris.

### 6. Install an Overflow System:

- Attach an overflow pipe to the tank to direct excess water away from the tank.

### 7. Set Up the Delivery System:

- Connect a pump or gravity-fed system to the tank outlet.

- Attach hoses or pipes to deliver water to your garden.

## Maintaining a Rainwater Harvesting System

Regular maintenance ensures that your rainwater harvesting system operates efficiently and provides clean water.

### 1. Clean Gutters and Downspouts:

- Regularly clean gutters and downspouts to remove leaves and debris.

## 2. Inspect the First Flush Diverter:

- ○ Check and clean the first flush diverter regularly to ensure it is functioning properly.

## 3. Clean the Filter:

- ○ Clean or replace the filter at the tank inlet as needed to maintain water quality.

## 4. Check the Storage Tank:

- ○ Inspect the tank for leaks, cracks, and algae growth.

- ○ Clean the tank periodically to remove sediment and debris.

## 5. Maintain the Delivery System:

- ○ Check the pump and hoses for clogs and leaks.

- ○ Test the system regularly to ensure it is delivering water efficiently.

# Combining Drip Irrigation and Rainwater Harvesting

Combining drip irrigation with rainwater harvesting creates a highly efficient and sustainable watering system. This integration maximizes water use efficiency and ensures your garden receives the best quality water.

## System Integration

1. **Design the Layout:**

   - Plan the layout to connect the rainwater storage tank to the drip irrigation system.

   - Position the tank at a higher elevation to utilize gravity flow or install a pump for water delivery.

2. **Install Connectors:**

   - Use appropriate connectors and fittings to link the storage tank outlet to the mainline tubing of the drip irrigation system.

   - Include a backflow preventer to prevent contamination of the storage tank.

3. **Use a Timer:**

   - Install a timer to automate watering schedules, ensuring consistent moisture levels in your garden.

   - Set the timer based on plant water needs and local weather conditions.

## Benefits of Combined Systems

1. **Water Conservation:**

   - Efficiently uses collected rainwater, reducing reliance on municipal water supplies.

2. **Improved Plant Health:**

○ Provides plants with high-quality, chemical-free rainwater.

### 3. Cost Savings:

○ Reduces water bills by utilizing free rainwater.

### 4. Sustainability:

○ Promotes sustainable gardening practices and reduces environmental impact.

## Advanced Water Management Techniques

For experienced gardeners and homesteaders, advanced water management techniques can further enhance water use efficiency and sustainability.

## Greywater Recycling

Greywater recycling involves reusing wastewater from sinks, showers, and washing machines for irrigation.

**Benefits:**

1. Conserves water by reusing household wastewater.

2. Reduces the load on septic systems and municipal wastewater treatment plants.

**System Components:**

1. **Greywater Collection:** Diverts greywater from household drains.

2. **Filtration:** Filters greywater to remove contaminants.

3. **Distribution:** Delivers filtered greywater to the garden through irrigation systems.

**Installation and Maintenance:**

1. Install a greywater system with proper plumbing and filtration.

2. Regularly maintain filters and check for clogs.

**Considerations:**

1. Use biodegradable soaps and detergents to avoid harmful chemicals.

2. Avoid using greywater on edible plants unless properly treated.

## Soil Moisture Sensors

Soil moisture sensors measure soil moisture levels and automate irrigation based on plant needs.

**Benefits:**

1. Optimizes watering schedules for efficient water use.

2. Prevents overwatering and underwatering.

**System Components:**

1. **Sensors:** Measure soil moisture at different depths.

2. **Controller:** Adjusts irrigation based on sensor readings.

**Installation and Maintenance:**

1. Install sensors at various locations and depths in the garden.

2. Regularly calibrate and maintain sensors for accurate readings.

## Water-Wise Landscaping

Water-wise landscaping involves designing gardens to reduce water use and improve drought tolerance.

1. **Benefits:**

   ○ Reduces water consumption.

   ○ Creates a low-maintenance and sustainable garden.

2. **Design Principles:**

   ○ **Plant Selection:** Choose drought-tolerant and native plants.

   ○ **Zoning:** Group plants with similar water needs together.

   ○ **Mulching:** Use mulch to retain soil moisture and reduce evaporation.

   ○ **Hardscaping:** Incorporate hardscape elements (rocks, gravel, pathways) to reduce irrigated areas.

Efficient water and irrigation management are essential for a productive and sustainable garden. By understanding plant water needs, implementing efficient watering techniques, and utilizing advanced systems like drip irrigation and rainwater harvesting, you can optimize water use and promote healthy plant growth. Combining these methods with sustainable practices such as greywater recycling and water-wise landscaping further enhances water efficiency and environmental stewardship. With careful planning and regular maintenance, you can create a thriving garden that conserves water and supports a resilient homestead.

# Chapter 12

# Home Canning and Preservation

Home canning and preservation are essential skills for any homesteader. These techniques allow you to store the bounty of your harvest for future use, ensuring that you have a supply of nutritious and delicious food throughout the year. This chapter will cover the basics of home canning, methods for preserving fruits and vegetables, and recipes and techniques for making jams, jellies, and pickles.

## Basics of Home Canning

Home canning is a method of preserving food in jars by using heat to destroy microorganisms that can cause spoilage. Proper canning techniques ensure food safety and maintain the quality and flavor of your preserved goods.

## Types of Canning

There are two primary methods of home canning: water bath canning and pressure canning. Each method is suitable for different types of foods based on their acidity levels.

## Water Bath Canning

Water bath canning is suitable for high-acid foods, which have a pH level of 4.6 or lower. The high acidity prevents the growth of bacteria, particularly Clostridium botulinum, which causes botulism.

1. **Suitable Foods:**

   - Fruits and fruit juices

   - Jams, jellies, and preserves

   - Pickles and relishes

   - Tomatoes with added acid (lemon juice or citric acid)

   - Vinegar-based sauces and condiments

2. **Equipment Needed:**

   - **Canning Jars:** Glass jars with two-piece lids (lid and band)

   - **Water Bath Canner:** Large pot with a lid and a rack

   - **Jar Lifter:** Tool for safely removing hot jars from the canner

   - **Lid Lifter:** Magnetic tool for lifting lids from hot water

   - **Funnel:** Wide-mouth funnel for filling jars

- **Bubble Remover/Headspace Tool:** Tool for removing air bubbles and measuring headspace

- **Kitchen Towels:** For cleaning jar rims and handling hot jars

## Pressure Canning

Pressure canning is required for low-acid foods, which have a pH level above 4.6. These foods must be processed at a higher temperature to destroy harmful bacteria.

1. **Suitable Foods:**

   - Vegetables (excluding tomatoes with added acid)

   - Meats and poultry

   - Seafood

   - Soups and stews

   - Beans and legumes

2. **Equipment Needed:**

   - **Canning Jars:** Glass jars with two-piece lids (lid and band)

   - **Pressure Canner:** Heavy-duty pot with a locking lid, pressure gauge, and a rack

   - **Jar Lifter:** Tool for safely removing hot jars from the canner

   - **Lid Lifter:** Magnetic tool for lifting lids from hot water

- ○ **Funnel:** Wide-mouth funnel for filling jars

- ○ **Bubble Remover/Headspace Tool:** Tool for removing air bubbles and measuring headspace

- ○ **Kitchen Towels:** For cleaning jar rims and handling hot jars

## Steps for Safe Canning

Regardless of the method used, following proper canning procedures is crucial to ensure food safety.

1. **Preparation:**

   - ○ **Wash Jars and Lids:** Wash jars, lids, and bands in hot, soapy water and rinse well. Keep jars hot until ready to use by placing them in a pot of hot water or in a warm oven. Follow the manufacturer's instructions for preparing lids.

   - ○ **Prepare Food:** Wash, peel, chop, or cook food according to the recipe. Fill jars with prepared food, leaving appropriate headspace (the space between the top of the food and the rim of the jar).

2. **Filling Jars:**

   - ○ **Add Food:** Use a funnel to fill jars with food, leaving the recommended headspace.

   - ○ **Remove Air Bubbles:** Run a bubble remover tool or a non-metallic spatula around the inside of the jar to release trapped air bubbles.

- **Clean Jar Rims:** Wipe the rims of the jars with a damp cloth to ensure a clean seal.

- **Apply Lids:** Place the lids on the jars and screw the bands on finger-tight.

### 3. Processing:

- **Water Bath Canning:** Place jars in the water bath canner, ensuring they are covered by at least 1-2 inches of water. Bring to a rolling boil and process for the time specified in the recipe.

- **Pressure Canning:** Place jars in the pressure canner, add the recommended amount of water, secure the lid, and heat according to the canner's instructions. Process for the time and pressure specified in the recipe.

### 4. Cooling and Sealing:

- **Remove Jars:** Use a jar lifter to carefully remove jars from the canner and place them on a towel or cooling rack.

- **Cool Jars:** Allow jars to cool undisturbed for 12-24 hours.

- **Check Seals:** After cooling, check the seals by pressing down on the center of each lid. If the lid does not flex, the jar is sealed. If the lid flexes, refrigerate and use the contents promptly or reprocess with a new lid.

### 5. Label and Store:

- **Label Jars:** Write the contents and date on the jar lids or use labels.

- **Store Jars:** Store sealed jars in a cool, dark, and dry place. Properly canned foods can last for up to one year.

## Preserving Fruits and Vegetables

Preserving fruits and vegetables through canning allows you to enjoy the flavors of your garden year-round. This section covers techniques and recipes for preserving a variety of fruits and vegetables.

## Preserving Fruits

Fruits can be preserved by canning them in syrup, juice, or water. Here are some popular methods and recipes for preserving fruits.

## Canning Fruit in Syrup

Canning fruit in syrup helps maintain the fruit's flavor, texture, and color. The syrup can be light, medium, or heavy, depending on your preference.

1. **Ingredients:**

   - Fresh fruit (peaches, pears, cherries, etc.)

   - Sugar

   - Water

2. **Syrup Preparation:**

   - **Light Syrup:** 2 cups sugar + 4 cups water

- **Medium Syrup:** 3 cups sugar + 4 cups water

- **Heavy Syrup:** 4 cups sugar + 4 cups water

- Combine sugar and water in a saucepan and bring to a boil, stirring until sugar is dissolved.

### 3. Canning Process:

- **Prepare Fruit:** Wash, peel, pit, and slice fruit as needed. Prevent browning by placing peeled fruit in a solution of water and lemon juice (1 tablespoon lemon juice per quart of water).

- **Pack Jars:** Pack fruit into hot jars, leaving 1/2 inch headspace. Pour hot syrup over fruit, maintaining the headspace.

- **Process Jars:** Process jars in a water bath canner for the recommended time based on the fruit type and jar size.

## Canning Fruit Juice

Canning fruit juice is a great way to preserve the fresh taste of fruits. Popular juices for canning include apple, grape, and berry juices.

### 1. Ingredients:

- Fresh fruit (apples, grapes, berries, etc.)

- Water (optional)

### 2. Juice Extraction:

- **Cold Press:** Use a fruit press or juicer to extract juice from the fruit.

- **Cooked Method:** Simmer fruit with a small amount of water until soft, then strain through a jelly bag or cheesecloth to extract juice.

3. **Canning Process:**

- **Heat Juice:** Heat juice to a simmer (190°F) without boiling.

- **Pack Jars:** Pour hot juice into hot jars, leaving 1/4 inch headspace.

- **Process Jars:** Process jars in a water bath canner for the recommended time based on the fruit type and jar size.

## Preserving Vegetables

Vegetables can be preserved through pressure canning, which ensures they are processed at a high enough temperature to destroy harmful bacteria. Here are some methods and recipes for preserving vegetables.

## Canning Vegetables

Pressure canning is essential for preserving low-acid vegetables to ensure food safety. Follow these general steps for canning vegetables.

1. **Preparation:**

- **Wash Vegetables:** Thoroughly wash vegetables to remove dirt and contaminants.

- **Prepare Vegetables:** Peel, trim, and cut vegetables as needed. Blanching (briefly boiling and then plunging into ice water) is recommended for most vegetables to preserve color, texture, and flavor.

## 2. Canning Process:

- **Pack Jars:** Pack prepared vegetables into hot jars, leaving 1-inch headspace. Add boiling water or broth to cover vegetables, maintaining the headspace.

- **Process Jars:** Process jars in a pressure canner for the recommended time and pressure based on the vegetable type and jar size.

# Canning Tomatoes

Tomatoes can be preserved using both water bath canning and pressure canning methods, provided they have added acid to ensure safe pH levels.

## 1. Ingredients:

- Fresh tomatoes

- Bottled lemon juice or citric acid

## 2. Preparation:

- **Wash and Peel Tomatoes:** Wash tomatoes and remove skins by blanching in boiling water for 30-60 seconds, then plunging into ice water. Slip off the skins and core tomatoes.

3. **Canning Process:**

- **Add Acid:** Add 1 tablespoon bottled lemon juice or 1/4 teaspoon citric acid per pint jar (2 tablespoons lemon juice or 1/2 teaspoon citric acid per quart jar).

- **Pack Jars:** Pack tomatoes into hot jars, leaving 1/2 inch headspace. Add boiling water, tomato juice, or leave them in their own juice.

- **Process Jars:** Process jars in a water bath canner (for high-acid tomatoes) or pressure canner (for low-acid varieties) for the recommended time and pressure.

# Making Jams, Jellies, and Pickles

Jams, jellies, and pickles are popular preserved foods that add flavor and variety to your pantry. This section covers the basics of making these delicious preserves.

## Making Jams and Jellies

Jams and jellies are made by cooking fruit with sugar and pectin to create a thick, spreadable product. The key difference is that jams include crushed fruit or fruit pulp, while jellies are made from fruit juice.

### Basic Jam Recipe

1. **Ingredients:**

- 4 cups crushed fruit (strawberries, raspberries, etc.)

- ○ 4 cups sugar

- ○ 1 package (1.75 oz) powdered pectin

- ○ 1/4 cup lemon juice (optional, for added acidity)

## 2. Preparation:

- ○ **Sterilize Jars:** Wash jars and lids, and sterilize them in boiling water for 10 minutes.

- ○ **Prepare Fruit:** Wash, hull, and crush fruit to measure the required amount.

## 3. Cooking Process:

- ○ **Combine Ingredients:** In a large pot, combine fruit, pectin, and lemon juice. Bring to a boil over high heat, stirring constantly.

- ○ **Add Sugar:** Add sugar all at once, stirring until dissolved. Return to a rolling boil and boil for 1-2 minutes.

- ○ **Test for Set:** Use the cold plate test (place a small amount of jam on a cold plate, let it cool for a few seconds, and then push it with your finger to see if it wrinkles and holds its shape) to check for the desired set.

## 4. Canning Process:

- ○ **Fill Jars:** Pour hot jam into hot jars, leaving 1/4 inch headspace. Wipe rims, apply lids, and screw bands on finger-tight.

- **Process Jars:** Process jars in a water bath canner for 10 minutes. Remove jars and cool for 12-24 hours. Check seals before storing.

## Basic Jelly Recipe

1. **Ingredients:**

   - 4 cups fruit juice (apple, grape, etc.)

   - 4 cups sugar

   - 1 package (1.75 oz) powdered pectin

   - 1/4 cup lemon juice (optional, for added acidity)

2. **Preparation:**

   - **Sterilize Jars:** Wash jars and lids, and sterilize them in boiling water for 10 minutes.

   - **Prepare Juice:** Extract juice from fruit by cooking it with a small amount of water, then straining through a jelly bag or cheesecloth.

3. **Cooking Process:**

   - **Combine Ingredients:** In a large pot, combine fruit juice, pectin, and lemon juice. Bring to a boil over high heat, stirring constantly.

   - **Add Sugar:** Add sugar all at once, stirring until dissolved.

Return to a rolling boil and boil for 1-2 minutes.

- ○ **Test for Set:** Use the cold plate test to check for the desired set.

### 4. Canning Process:

- ○ **Fill Jars:** Pour hot jelly into hot jars, leaving 1/4 inch headspace. Wipe rims, apply lids, and screw bands on finger-tight.

- ○ **Process Jars:** Process jars in a water bath canner for 10 minutes. Remove jars and cool for 12-24 hours. Check seals before storing.

## Making Pickles

Pickling is a preservation method that uses vinegar, salt, and spices to preserve vegetables. The high acidity of vinegar ensures food safety and enhances flavor.

### Basic Pickle Recipe

#### 1. Ingredients:

- ○ 4 cups cucumbers, sliced or whole

- ○ 2 cups water

- ○ 2 cups vinegar (5% acidity)

- ○ 1/4 cup canning salt

- ○ 1 tablespoon pickling spice

- Fresh dill (optional)

- Garlic cloves (optional)

2. **Preparation:**

- **Sterilize Jars:** Wash jars and lids, and sterilize them in boiling water for 10 minutes.

- **Prepare Cucumbers:** Wash cucumbers and trim ends. Slice or leave whole as desired.

3. **Brine Preparation:**

- **Combine Ingredients:** In a large pot, combine water, vinegar, salt, and pickling spice. Bring to a boil, stirring until salt is dissolved.

4. **Canning Process:**

- **Pack Jars:** Pack cucumbers into hot jars, adding dill and garlic if desired. Pour hot brine over cucumbers, leaving 1/2 inch headspace.

- **Process Jars:** Process jars in a water bath canner for 10 minutes. Remove jars and cool for 12-24 hours. Check seals before storing.

## Fermented Pickles

Fermented pickles, also known as lacto-fermented pickles, use natural fermentation to develop flavor and preserve vegetables.

1. **Ingredients:**

   - 4 cups cucumbers, sliced or whole

   - 4 cups water

   - 2 tablespoons canning salt

   - Fresh dill (optional)

   - Garlic cloves (optional)

   - Mustard seeds, peppercorns, or other spices (optional)

2. **Preparation:**

   - **Sterilize Jars:** Wash jars and lids, and sterilize them in boiling water for 10 minutes.

   - **Prepare Cucumbers:** Wash cucumbers and trim ends. Slice or leave whole as desired.

3. **Brine Preparation:**

   - **Combine Ingredients:** In a large pot, dissolve salt in water to make a brine.

4. **Fermentation Process:**

   - **Pack Jars:** Pack cucumbers into hot jars, adding dill, garlic, and spices if desired. Pour brine over cucumbers, leaving 1 inch headspace.

   - **Ferment:** Cover jars with a fermentation lid or cloth to allow gases to escape. Let jars sit at room temperature for 1-4 weeks, depending on desired flavor. Check daily for mold and remove if present.

○ **Store:** Once fermented to your liking, transfer jars to the refrigerator to slow fermentation and store for several months.

Home canning and preservation are invaluable skills for extending the harvest and ensuring a supply of healthy, homegrown food throughout the year. By mastering the basics of canning, preserving fruits and vegetables, and making jams, jellies, and pickles, you can enjoy the flavors of your garden long after the growing season has ended. With attention to detail and proper techniques, home canning and preservation can be a safe, rewarding, and sustainable practice that enhances your homesteading experience.

# Chapter 13

# Beekeeping for Beginners

Beekeeping is a rewarding and fascinating hobby that provides numerous benefits to a homestead. Honeybees play a crucial role in pollination, which boosts the productivity of gardens and orchards. Additionally, beekeeping offers the sweet reward of honey and other bee products. This chapter will cover the importance of bees in homesteading, how to set up your beehive, and the essentials of caring for bees and harvesting honey.

## Importance of Bees in Homesteading

Bees, particularly honeybees, are vital to the ecosystem and agriculture. Their role extends far beyond honey production, encompassing pollination and maintaining biodiversity.

## Pollination

Pollination is the process by which bees transfer pollen from one flower to another, facilitating plant reproduction. This process is essential for the production of fruits, vegetables, and seeds.

1. **Crop Yields:**

   ○ **Enhanced Productivity:** Bees significantly increase the yields of many crops, including apples, blueberries, cucumbers, and almonds.

   ○ **Quality of Produce:** Pollination by bees can improve the size, shape, and quality of fruits and vegetables.

2. **Biodiversity:**

   ○ **Wild Plants:** Bees pollinate many wild plants, contributing to biodiversity and the health of natural ecosystems.

   ○ **Habitat Creation:** The presence of a diverse range of plants provides habitats for other wildlife, supporting a balanced ecosystem.

## Honey Production

Honey is the most well-known product of beekeeping. It is a versatile natural sweetener with numerous culinary and medicinal uses.

1. **Nutritional Value:**

   ○ **Natural Sweetener:** Honey is a healthier alternative to refined sugars, rich in antioxidants, enzymes, and minerals.

- **Energy Source:** It provides a quick source of energy due to its simple sugars, glucose and fructose.

## 2. Medicinal Uses:

- **Wound Healing:** Honey has antibacterial properties and can be used for wound healing and treating burns.

- **Cough Suppressant:** It is effective in soothing sore throats and reducing cough symptoms.

## 3. Culinary Uses:

- **Cooking and Baking:** Honey can be used in a variety of recipes, including baked goods, marinades, and dressings.

- **Preservation:** It acts as a natural preservative due to its low moisture content and acidic pH.

## Other Bee Products

In addition to honey, bees produce several other valuable products.

## 1. Beeswax:

- **Uses:** Beeswax is used in making candles, cosmetics, and polishes. It is also used in food wraps and as a coating for cheese.

- **Properties:** It is a natural emulsifier and has moisturizing properties, making it ideal for skin care products.

## 2. Propolis:

- **Uses:** Propolis, a resinous substance collected by bees from tree buds, is used in health supplements and natural remedies due to its antimicrobial properties.

- **Applications:** It can be used to treat minor wounds, sore throats, and infections.

3. **Royal Jelly:**

- **Uses:** Royal jelly, a secretion used to feed larvae and the queen bee, is a popular dietary supplement believed to have numerous health benefits.

- **Properties:** It is rich in vitamins, minerals, and amino acids, and is used in anti-aging products and health supplements.

4. **Pollen:**

- **Uses:** Bee pollen is harvested and used as a dietary supplement. It is rich in proteins, vitamins, and minerals.

- **Applications:** It can be added to smoothies, yogurt, and other foods for a nutritional boost.

## Setting Up Your Beehive

Setting up a beehive involves selecting the right equipment, choosing a suitable location, and acquiring bees. Proper setup is essential for the health and productivity of your hive.

## Selecting the Right Equipment

Several types of hives and tools are available for beekeeping. Here is an overview of the essential equipment you will need.

1. **Hive Types:**

   ○ **Langstroth Hive:** The most common type, consisting of stacked rectangular boxes (supers) with removable frames. It is easy to manage and provides high honey yields.

   ○ **Top-Bar Hive:** A horizontal hive with bars across the top for bees to build their comb. It is simpler to construct and manage, but honey yields are typically lower.

   ○ **Warre Hive:** A vertical hive with boxes added to the bottom. It mimics the natural environment of bees and requires minimal intervention.

2. **Hive Components:**

   ○ **Bottom Board:** The base of the hive, which supports the hive structure and provides an entrance for bees.

   ○ **Hive Bodies (Supers):** Boxes that hold frames where bees build their comb. Deep supers are used for brood, and shallow supers are used for honey.

   ○ **Frames:** Removable frames that hold the comb. They can be equipped with foundation to guide comb building.

   ○ **Inner Cover:** A cover placed over the top super to provide insulation and ventilation.

- **Outer Cover:** A weatherproof cover that protects the hive from the elements.

3. **Beekeeping Tools:**

   - **Hive Tool:** A metal tool used to pry apart hive components and scrape off excess propolis and wax.

   - **Smoker:** A device that produces smoke to calm bees and reduce their defensive behavior.

   - **Bee Brush:** A soft brush used to gently remove bees from frames during inspections.

   - **Feeder:** A container used to provide supplemental feed (sugar syrup) to bees, especially during periods of low nectar flow.

4. **Protective Gear:**

   - **Bee Suit:** A full-body suit that protects against bee stings.

   - **Gloves:** Long, thick gloves to protect hands and wrists.

   - **Veil:** A mesh hat and veil to protect the face and neck.

## Choosing a Suitable Location

Selecting the right location for your beehive is crucial for the well-being of your bees and the success of your beekeeping venture.

1. **Sunlight:**

   - **Morning Sun:** Place the hive where it receives morning sunlight to encourage bees to start foraging early.

- ◦ **Partial Shade:** Ensure the hive receives some shade during the hottest part of the day to prevent overheating.

## 2. **Wind Protection:**

- ◦ **Sheltered Location:** Place the hive in a sheltered area to protect it from strong winds. Use natural windbreaks like hedges or fences if necessary.

## 3. **Water Source:**

- ◦ **Nearby Water:** Ensure there is a reliable water source nearby, such as a pond, stream, or birdbath. Bees need water for hydration and to regulate hive temperature.

## 4. **Foraging Area:**

- ◦ **Abundant Flowers:** Choose a location with abundant flowering plants within a 2-3 mile radius to provide a steady supply of nectar and pollen.

- ◦ **Diverse Flora:** A variety of flowering plants will ensure a diverse diet for the bees.

## 5. **Accessibility:**

- ◦ **Ease of Access:** Place the hive in an area that is easily accessible for inspections and maintenance.

- ◦ **Safety:** Ensure the hive is placed away from high-traffic areas, children, and pets to avoid disturbances and stings.

## 6. Legal Considerations:

- **Local Regulations:** Check local zoning laws and regulations regarding beekeeping. Some areas have restrictions on hive placement and the number of hives allowed.

# Acquiring Bees

You can acquire bees through several methods, each with its own advantages.

## 1. Package Bees:

- **Description:** A package of bees consists of approximately 3 pounds of bees (about 10,000 bees) and a queen in a separate cage.

- **Advantages:** Easier to transport and install. Suitable for starting a new hive.

- **Source:** Purchase from reputable suppliers or beekeeping associations.

## 2. Nucleus Colony (Nuc):

- **Description:** A nuc is a small, established colony with 4-5 frames of brood, honey, and pollen, along with a queen.

- **Advantages:** Faster colony growth due to the presence of brood and a functioning queen.

- **Source:** Purchase from local beekeepers or beekeeping associations.

3. **Swarm:**

- ○ **Description:** A swarm is a natural group of bees that has left an established colony to form a new one.

- ○ **Advantages:** Free and a natural way to acquire bees. Requires knowledge and experience to capture and hive a swarm.

- ○ **Source:** Monitor for swarms in your area during spring and early summer.

4. **Established Colony:**

- ○ **Description:** An established colony is a fully functioning hive with brood, honey, and a queen.

- ○ **Advantages:** Immediate honey production and established colony dynamics.

- ○ **Source:** Purchase from beekeepers looking to sell or split their colonies.

## Caring for Bees and Harvesting Honey

Proper care and management are essential for maintaining a healthy and productive beehive. Regular inspections, monitoring for diseases and pests, and timely interventions will ensure the well-being of your bees.

## Regular Hive Inspections

Regular inspections help you monitor the health and productivity of your hive. Inspections should be conducted every 7-10 days during the active season (spring and summer).

1. **Preparation:**

   ○ **Gather Tools:** Ensure you have all necessary tools, including the hive tool, smoker, bee brush, and protective gear.

   ○ **Calm Bees:** Light the smoker and use a few puffs of smoke at the hive entrance and under the lid to calm the bees.

2. **Inspection Process:**

   ○ **Open the Hive:** Carefully remove the outer cover and inner cover using the hive tool.

   ○ **Check for Queen:** Look for the queen and check for signs of her presence, such as eggs and young larvae.

   ○ **Assess Brood:** Examine the brood pattern. Healthy brood should be evenly distributed and free of gaps.

   ○ **Check for Pests and Diseases:** Look for signs of pests such as varroa mites, small hive beetles, and wax moths. Inspect for symptoms of diseases like American foulbrood, European foulbrood, and chalkbrood.

   ○ **Monitor Honey Stores:** Check the honey stores to ensure the colony has enough food. Supplement with sugar syrup if necessary.

## 3. Record Keeping:

- **Maintain Records:** Keep detailed records of each inspection, noting the hive's condition, queen status, brood pattern, pest, and disease presence, and honey stores.

- **Track Progress:** Use the records to track the colony's progress and identify trends or recurring issues.

## Managing Pests and Diseases

Bees are susceptible to various pests and diseases that can weaken or destroy a colony. Effective management and preventive measures are crucial.

### 1. Varroa Mites:

- **Description:** Varroa mites are parasitic mites that feed on bee larvae and adults, weakening the colony and spreading diseases.

- **Detection:** Use methods like the sugar shake, alcohol wash, or sticky board test to monitor mite levels.

- **Treatment:** Implement treatments such as powdered sugar dusting, essential oil treatments (thymol, oxalic acid), or commercially available miticides.

### 2. Small Hive Beetles:

- **Description:** Small hive beetles (SHB) are pests that infest hives, laying eggs that hatch into larvae that feed on honey and pollen.

- **Detection:** Monitor for beetles and larvae during hive

inspections.

- **Management:** Use traps, reduce hive entrances, and maintain strong, healthy colonies to deter infestations.

3. **Wax Moths:**

- **Description:** Wax moths lay eggs in hives, and the larvae tunnel through comb, causing damage.

- **Detection:** Look for tunnels in comb and webbing in the hive.

- **Management:** Maintain strong colonies and store empty comb in a freezer to prevent infestations.

4. **American Foulbrood (AFB):**

- **Description:** AFB is a highly contagious bacterial disease that affects bee larvae.

- **Symptoms:** Sunken, darkened brood caps, foul smell, and "ropey" larval remains.

- **Management:** Destroy infected colonies and equipment. Use antibiotics only as a last resort and according to local regulations.

5. **European Foulbrood (EFB):**

- **Description:** EFB is a bacterial disease that affects bee larvae.

- **Symptoms:** Discolored larvae, twisted in their cells, and a sour smell.

- **Management:** Requeen the colony and provide supplemental feeding. Antibiotics may be used as directed.

6. **Nosema:**

- **Description:** Nosema is a fungal disease that affects the digestive tract of bees.

- **Symptoms:** Dysentery, reduced foraging activity, and increased winter mortality.

- **Management:** Improve hive ventilation and provide a healthy diet. Use treatments like fumagillin if necessary.

## Seasonal Management

Managing your beehive requires different strategies throughout the year to ensure the health and productivity of your bees.

1. **Spring:**

- **Hive Inspections:** Begin regular inspections to assess colony health and population growth.

- **Swarm Prevention:** Monitor for signs of swarming, such as queen cells and crowded conditions. Split hives or provide additional space if necessary.

- **Supplemental Feeding:** Provide sugar syrup if nectar sources are limited.

2. **Summer:**

- **Honey Production:** Add supers as needed to accommodate honey production.

- **Pest and Disease Management:** Monitor for pests and diseases and implement treatments as needed.

- **Queen Management:** Ensure the queen is healthy and laying eggs. Requeen if necessary.

3. **Fall:**

- **Honey Harvest:** Harvest honey before the end of the nectar flow, leaving enough for the bees to overwinter.

- **Winter Preparations:** Reduce hive entrances, insulate hives, and ensure adequate ventilation. Provide supplemental feeding if honey stores are low.

4. **Winter:**

- **Minimal Disturbance:** Avoid disturbing the hive during cold weather. Check periodically for signs of life and adequate food stores.

- **Supplemental Feeding:** Provide sugar blocks or fondant if honey stores are depleted.

## Harvesting Honey

Harvesting honey is one of the most rewarding aspects of beekeeping. Proper techniques ensure you collect high-quality honey without harming the bees.

## 1. Timing:

- **Mature Honey:** Harvest honey when at least 80% of the cells in the frames are capped. This indicates that the honey is fully ripened and has the right moisture content.

- **End of Nectar Flow:** Harvest honey at the end of the nectar flow to ensure the bees have sufficient food stores.

## 2. Equipment:

- **Honey Extractor:** A manual or electric device that uses centrifugal force to extract honey from the comb.

- **Uncapping Knife:** A tool used to remove the wax cappings from the honeycomb.

- **Strainer:** A fine mesh strainer to filter out wax particles and debris from the extracted honey.

- **Honey Bottles:** Clean, food-grade containers for storing the honey.

## 3. Harvesting Process:

- **Remove Frames:** Gently remove honey-filled frames from the hive, using a bee brush or escape board to clear bees from the frames.

- **Uncap Frames:** Use an uncapping knife to slice off the wax cappings from the honeycomb cells.

- **Extract Honey:** Place the frames in the honey extractor and spin to remove honey from the comb. Collect the honey in a clean container.

- **Strain Honey:** Pour the extracted honey through a fine mesh strainer to remove impurities.

- **Bottle Honey:** Transfer the strained honey into clean bottles or jars for storage.

4. **Storage:**

- **Proper Conditions:** Store honey in a cool, dry place away from direct sunlight. Honey should be kept in tightly sealed containers to prevent moisture absorption and fermentation.

- **Shelf Life:** Properly stored honey can last indefinitely, though it may crystallize over time. Crystallized honey can be liquefied by gently warming the container in a water bath.

Beekeeping is a fulfilling and valuable addition to any homestead. By understanding the importance of bees in pollination and honey production, setting up a well-designed beehive, and providing proper care and management, you can enjoy the numerous benefits of beekeeping. Regular inspections, effective pest and disease management, and seasonal adjustments are essential for maintaining healthy and productive hives. Harvesting honey and other bee products allows you to savor the fruits of your labor while supporting a sustainable and thriving ecosystem. With dedication and attention to detail, beekeeping can become a rewarding and enjoyable part of your homesteading journey.

# Chapter 14

# Raising Other Livestock

Raising livestock on a homestead can provide a sustainable source of meat, milk, wool, and other products. This chapter will focus on raising goats, sheep, and rabbits, covering the selection of appropriate breeds, the necessary housing and feeding requirements, and proper care techniques.

## Goats

Goats are versatile animals that can provide milk, meat, fiber, and even weed control. They are relatively easy to manage and can adapt to various climates.

## Choosing the Right Breeds

When choosing goat breeds, consider your goals, such as milk production, meat, fiber, or land management.

### Dairy Goats

1. **Nubian:**

   - **Milk Production:** High butterfat content (4-5%) ideal for making cheese and butter.

- ○ **Temperament:** Friendly, social, and vocal.

- ○ **Description:** Long ears, Roman nose, and a variety of colors and patterns.

## 2. Alpine:

- ○ **Milk Production:** High milk yield with lower butterfat content (3-4%).

- ○ **Temperament:** Energetic and curious.

- ○ **Description:** Medium to large size, erect ears, and various color patterns.

## 3. Saanen:

- ○ **Milk Production:** High milk yield with moderate butterfat content (3-4%).

- ○ **Temperament:** Calm and gentle.

- ○ **Description:** Large size, white or cream coat, and erect ears.

# Meat Goats

## 1. Boer:

- ○ **Growth Rate:** Fast-growing and high meat yield.

- ○ **Temperament:** Docile and easy to handle.

- **Description:** White body with a red head, muscular build.

2. **Kiko:**

- **Hardiness:** Resistant to parasites and diseases, good for low-input systems.

- **Temperament:** Independent and hardy.

- **Description:** Various colors, typically white or cream, with a sturdy build.

## Fiber Goats

1. **Angora:**

- **Fiber Production:** Produces mohair, a silky and lustrous fiber.

- **Temperament:** Gentle but require more care.

- **Description:** Long, curly coat, typically white, and medium size.

2. **Cashmere:**

- **Fiber Production:** Produces cashmere, a fine and soft fiber.

- **Temperament:** Hardy and adaptable.

- **Description:** Various colors, often dual-purpose for meat and fiber.

## Land Management

1. **Pygmy:**

   ○ **Grazing:** Efficient browsers for weed control.

   ○ **Temperament:** Friendly and good for small homesteads.

   ○ **Description:** Small size, various colors, and compact build.

2. **Nigerian Dwarf:**

   ○ **Grazing:** Effective foragers and also good milk producers.

   ○ **Temperament:** Gentle and playful.

   ○ **Description:** Small size, various colors, and upright ears.

## Housing

Goats need secure, comfortable housing to protect them from predators and harsh weather. Proper housing also helps manage health and productivity.

1. **Shelter Requirements:**

   ○ **Space:** Provide at least 20 square feet per goat in the shelter.

   ○ **Ventilation:** Ensure good airflow to reduce respiratory problems.

   ○ **Bedding:** Use straw, wood shavings, or hay for bedding, and keep it clean and dry.

2. **Fencing:**

   ○ **Type:** Use strong, secure fencing such as woven wire or electric fencing to prevent escapes.

- **Height:** Fencing should be at least 4-5 feet high.

- **Maintenance:** Regularly inspect and repair fencing to ensure its integrity.

3. **Pasture Management:**

- **Rotation:** Rotate pastures to prevent overgrazing and parasite buildup.

- **Forage:** Provide a mix of grasses, legumes, and browse plants.

## Feeding

Proper nutrition is crucial for the health and productivity of goats.

1. **Diet:**

- **Forage:** Goats thrive on a diet of pasture, hay, and browse.

- **Concentrates:** Supplement with grains if necessary, especially for lactating or growing goats.

- **Minerals:** Provide a mineral supplement formulated for goats, including salt and essential vitamins and minerals.

2. **Water:**

- **Availability:** Ensure goats have constant access to clean, fresh water.

- **Location:** Place water sources in shaded areas to keep the water cool and prevent contamination.

3. **Feeding Management:**

- **Hay Racks:** Use hay racks to minimize waste and keep feed clean.

- **Feeding Schedule:** Feed goats twice a day, adjusting the amount based on their needs and condition.

## Care

Regular care and maintenance are essential for keeping goats healthy.

1. **Health Checks:**

- **Frequency:** Conduct health checks at least once a month.

- **Vital Signs:** Monitor weight, body condition, coat quality, and behavior.

- **Signs of Illness:** Watch for signs of illness such as lethargy, coughing, diarrhea, or changes in appetite.

2. **Parasite Control:**

- **Deworming:** Implement a strategic deworming program based on fecal egg counts and veterinary advice.

- **Pasture Management:** Rotate pastures and avoid overgrazing to reduce parasite loads.

### 3. Hoof Trimming:

- **Frequency:** Trim hooves every 6-8 weeks to prevent overgrowth and lameness.

- **Tools:** Use sharp hoof trimmers and clean the hooves thoroughly before trimming.

### 4. Vaccinations:

- **Core Vaccines:** Administer vaccines for common diseases such as tetanus, enterotoxemia, and pneumonia.

- **Schedule:** Follow a vaccination schedule recommended by a veterinarian.

# Sheep

Sheep are valued for their wool, meat, and milk. They are relatively easy to manage and can be an excellent addition to a homestead.

## Choosing the Right Breeds

Selecting the appropriate breed of sheep depends on your goals, whether for wool, meat, or milk production.

## Wool Sheep

### 1. Merino:

- **Wool Quality:** Fine, high-quality wool with a high yield.

- Temperament: Hardy and adaptable.

- Description: White wool, medium to large size.

## 2. Suffolk:

- Wool Quality: Medium wool, also good for meat production.

- Temperament: Energetic and hardy.

- Description: Black face and legs, large size.

## Meat Sheep

### 1. Dorper:

- Growth Rate: Fast-growing with excellent meat quality.

- Temperament: Hardy and low-maintenance.

- Description: White body with a black head, muscular build.

### 2. Hampshire:

- Growth Rate: Rapid growth and high meat yield.

- Temperament: Calm and easy to handle.

- Description: White body with a black face and legs, large size.

## Dairy Sheep

### 1. East Friesian:

- **Milk Production:** High milk yield, ideal for cheese making.

- **Temperament:** Friendly and docile.

- **Description:** White wool, medium to large size.

## 2. Lacaune:

- **Milk Production:** Good milk yield with high butterfat content.

- **Temperament:** Hardy and adaptable.

- **Description:** White wool, medium size.

# Housing

Proper housing is essential for the health and productivity of sheep.

## 1. Shelter Requirements:

- **Space:** Provide at least 20-25 square feet per sheep in the shelter.

- **Ventilation:** Ensure good airflow to reduce respiratory problems.

- **Bedding:** Use straw or wood shavings for bedding, and keep it clean and dry.

## 2. Fencing:

- **Type:** Use strong, secure fencing such as woven wire or electric fencing to prevent escapes.

- ◦ **Height:** Fencing should be at least 4 feet high.

- ◦ **Maintenance:** Regularly inspect and repair fencing to ensure its integrity.

### 3. Pasture Management:

- ◦ **Rotation:** Rotate pastures to prevent overgrazing and parasite buildup.

- ◦ **Forage:** Provide a mix of grasses and legumes.

## Feeding

Proper nutrition is crucial for the health and productivity of sheep.

### 1. Diet:

- ◦ **Forage:** Sheep thrive on a diet of pasture, hay, and browse.

- ◦ **Concentrates:** Supplement with grains if necessary, especially for lactating or growing sheep.

- ◦ **Minerals:** Provide a mineral supplement formulated for sheep, including salt and essential vitamins and minerals.

### 2. Water:

- ◦ **Availability:** Ensure sheep have constant access to clean, fresh water.

- ◦ **Location:** Place water sources in shaded areas to keep the water cool and prevent contamination.

### 3. Feeding Management:

- **Hay Racks:** Use hay racks to minimize waste and keep feed clean.

- **Feeding Schedule:** Feed sheep twice a day, adjusting the amount based on their needs and condition.

## Care

Regular care and maintenance are essential for keeping sheep healthy.

### 1. Health Checks:

- **Frequency:** Conduct health checks at least once a month.

- **Vital Signs:** Monitor weight, body condition, coat quality, and behavior.

- **Signs of Illness:** Watch for signs of illness such as lethargy, coughing, diarrhea, or changes in appetite.

### 2. Parasite Control:

- **Deworming:** Implement a strategic deworming program based on fecal egg counts and veterinary advice.

- **Pasture Management:** Rotate pastures and avoid overgrazing to reduce parasite loads.

### 3. Hoof Trimming:

- **Frequency:** Trim hooves every 6-8 weeks to prevent overgrowth and lameness.

- ○ **Tools:** Use sharp hoof trimmers and clean the hooves thoroughly before trimming.

4. **Shearing:**

- ○ **Frequency:** Shear wool at least once a year, preferably in spring.

- ○ **Tools:** Use electric or hand shears, and ensure the sheep is clean before shearing.

5. **Vaccinations:**

- ○ **Core Vaccines:** Administer vaccines for common diseases such as tetanus, enterotoxemia, and pneumonia.

- ○ **Schedule:** Follow a vaccination schedule recommended by a veterinarian.

# Rabbits

Rabbits are excellent for small-scale meat production and can also provide fur and manure for the garden. They are easy to raise and require minimal space.

## Choosing the Right Breeds

Selecting the appropriate rabbit breed depends on your goals, whether for meat, fur, or pet purposes.

## Meat Rabbits

### 1. New Zealand:

- **Growth Rate:** Fast-growing with high meat yield.

- **Temperament:** Calm and easy to handle.

- **Description:** White, red, or black fur, medium to large size.

### 2. Californian:

- **Growth Rate:** Rapid growth and excellent meat quality.

- **Temperament:** Friendly and docile.

- **Description:** White fur with black points (ears, nose, feet, and tail), medium to large size.

## Fur Rabbits

### 1. Rex:

- **Fur Quality:** Dense, plush fur ideal for fur production.

- **Temperament:** Friendly and gentle.

- **Description:** Various colors, medium size.

### 2. Angora:

- **Fur Quality:** Long, silky wool used for spinning and textiles.

- **Temperament:** Calm but require regular grooming.

- **Description:** Various colors, medium size.

## Pet Rabbits

### 1. Holland Lop:

- **Size:** Small size with a compact build.

- **Temperament:** Friendly and playful.

- **Description:** Lop ears and various colors.

### 2. Mini Rex:

- **Size:** Small size with a dense, plush coat.

- **Temperament:** Gentle and easy to handle.

- **Description:** Various colors and patterns.

## Housing

Proper housing is essential for the health and productivity of rabbits.

### 1. Hutch Requirements:

- **Space:** Provide at least 4 square feet per rabbit in the hutch.

- **Ventilation:** Ensure good airflow to reduce respiratory problems.

- **Bedding:** Use straw, wood shavings, or hay for bedding, and keep it clean and dry.

2. **Cages:**

- ○ **Type:** Use wire cages with a solid floor or a tray to collect waste.

- ○ **Size:** Provide enough space for the rabbit to move around comfortably.

- ○ **Location:** Place cages in a sheltered area, protected from direct sunlight, wind, and rain.

3. **Pasture Management:**

- ○ **Rotation:** If raising rabbits on pasture, use movable pens to rotate grazing areas.

- ○ **Forage:** Provide a mix of grasses and legumes for grazing.

## Feeding

Proper nutrition is crucial for the health and productivity of rabbits.

1. **Diet:**

- ○ **Hay:** Provide unlimited access to high-quality hay, such as timothy or alfalfa, to support digestion and dental health.

- ○ **Pellets:** Supplement with a commercial rabbit pellet formulated to meet nutritional needs.

- ○ **Fresh Vegetables:** Offer a variety of fresh vegetables, such as leafy greens, carrots, and bell peppers, as treats.

2. **Water:**

- **Availability:** Ensure rabbits have constant access to clean, fresh water.

- **Location:** Use water bottles or bowls, and check frequently to ensure they are not empty or contaminated.

### 3. Feeding Management:

- **Hay Racks:** Use hay racks to minimize waste and keep feed clean.

- **Feeding Schedule:** Feed rabbits twice a day, adjusting the amount based on their needs and condition.

# Care

Regular care and maintenance are essential for keeping rabbits healthy.

### 1. Health Checks:

- **Frequency:** Conduct health checks at least once a month.

- **Vital Signs:** Monitor weight, body condition, coat quality, and behavior.

- **Signs of Illness:** Watch for signs of illness such as lethargy, coughing, diarrhea, or changes in appetite.

### 2. Parasite Control:

- **Deworming:** Implement a strategic deworming program based on veterinary advice.

- **Cage Cleaning:** Clean cages regularly to prevent the buildup of waste and reduce the risk of parasites.

## 3. Nail Trimming:

- **Frequency:** Trim nails every 4-6 weeks to prevent overgrowth and injury.

- **Tools:** Use sharp nail clippers and handle rabbits gently during trimming.

## 4. Grooming:

- **Frequency:** Regularly groom rabbits, especially long-haired breeds, to prevent matting and hairballs.

- **Tools:** Use grooming brushes and combs suitable for the rabbit's coat type.

## 5. Breeding:

- **Selection:** Choose healthy, compatible rabbits for breeding.

- **Gestation:** Monitor the doe (female rabbit) during the 30-31 day gestation period.

- **Kits (Baby Rabbits):** Provide a nesting box for the doe to give birth. Ensure kits are warm and safe, and monitor their growth.

Raising goats, sheep, and rabbits on a homestead can provide a sustainable source of milk, meat, fiber, and other products. By choosing the right breeds, providing proper housing and nutrition, and ensuring regular care and maintenance, you can successfully manage these livestock and enjoy the benefits they offer. With attention to detail and dedication, raising livestock can enhance your homesteading experience and contribute to a self-sufficient lifestyle.

# Chapter 15

# Maximizing Your Yield

Homesteading involves not only growing and harvesting crops but also ensuring that you make the most of your garden's potential. Maximizing your yield requires extending the growing season, utilizing various structures to protect plants, and employing advanced gardening techniques to continue producing crops throughout fall and winter. In this chapter, we'll explore ways to extend the growing season, the use of cold frames and greenhouses, fall and winter gardening, and other season extension techniques.

## Extending the Growing Season

One of the key strategies to maximize your garden's yield is to extend the growing season. This involves creating conditions that allow plants to thrive beyond their typical growing period. By protecting plants from frost and providing them with optimal growing conditions, you can harvest crops for a longer period.

The first step in extending your growing season is understanding your local climate and frost dates. This information will help you plan when to start seeds indoors, when to transplant seedlings outside, and when to implement protective measures against cold weather.

Start by investing in a soil thermometer. Soil temperature is crucial for germination and root development. Warm soil encourages seeds to germinate quickly and grow robustly. For early spring planting, you can warm the soil by covering it with black plastic or using cloches. These simple techniques can raise the soil temperature by several degrees, allowing you to plant earlier than usual.

Transplanting seedlings into the garden earlier can be achieved by hardening them off properly. Hardening off is the process of gradually acclimating indoor-grown plants to outdoor conditions. Start by placing seedlings outside for a few hours a day, gradually increasing their exposure to sunlight and outdoor temperatures over a week or two. This process strengthens the plants, reducing transplant shock and increasing their chances of thriving.

Once your plants are in the ground, protecting them from unexpected frosts becomes crucial. Frost cloths, also known as row covers, are an excellent investment. These lightweight fabrics can be draped over plants to protect them from frost while still allowing light and moisture to penetrate. Make sure to anchor the edges of the frost cloth securely to prevent it from blowing away.

Watering practices also play a significant role in extending the growing season. Well-watered soil retains more heat than dry soil, providing a buffer against cold temperatures. Water your garden thoroughly in the afternoon before an expected frost. The water will release heat slowly during the night, helping to protect the roots of your plants.

In addition to these basic strategies, more advanced techniques such as using cold frames and greenhouses can significantly extend your growing season and increase your garden's productivity.

## Using Cold Frames and Greenhouses

Cold frames and greenhouses are invaluable tools for any serious gardener. They provide controlled environments that protect plants from extreme weather, allowing for earlier planting in the spring and extended harvests into the fall and winter.

## Cold Frames

Cold frames are simple structures consisting of a transparent cover, usually made of glass or clear plastic, mounted over a wooden or metal frame. They act as mini-greenhouses, capturing solar energy and creating a microclimate that is warmer than the surrounding air.

To build a cold frame, you can repurpose old windows or use clear plastic sheeting stretched over a wooden frame. Place the frame on a south-facing slope to maximize sunlight exposure. The back of the frame should be higher than the front to allow the cover to slant towards the sun, enhancing heat absorption and promoting water drainage.

Cold frames are ideal for hardening off seedlings, starting cool-season crops earlier in the spring, and extending the harvest of fall crops. In the spring, use the cold frame to start lettuce, spinach, radishes, and other cold-tolerant vegetables. As the weather warms, the cold frame can be used to transition heat-loving plants like tomatoes and peppers from the indoors to the garden.

During fall and winter, cold frames can protect crops from frost and wind, allowing you to harvest fresh greens and root vegetables long after the first frost. Insulate the cold frame with straw or hay bales around the exterior during extreme cold snaps, and cover the top with an additional layer of burlap or an old blanket at night to retain heat.

## Greenhouses

Greenhouses are more permanent structures that provide a controlled environment for growing plants year-round. They come in various sizes and designs, from small, simple hoop houses to large, elaborate glass structures. The type of greenhouse you choose will depend on your budget, space, and gardening goals.

A basic hoop house is an affordable and effective option for extending your growing season. It consists of a series of hoops made from PVC or metal, covered with clear plastic sheeting. Hoop houses can be erected quickly and provide excellent protection against frost, wind, and pests. They are ideal for growing a wide variety of vegetables and herbs throughout the year.

For those with a larger budget and more ambitious gardening plans, investing in a more substantial greenhouse can provide numerous benefits. A well-constructed greenhouse allows for precise control of temperature, humidity, and light, enabling you to grow a diverse range of plants, including those that would not typically thrive in your climate.

Inside the greenhouse, raised beds or benches can be used to optimize space and improve plant health. Raised beds provide better drainage and soil quality, while benches allow for easy access and efficient use of space. Install shelving and hanging racks to maximize vertical space for starting seedlings or growing smaller plants.

To maintain optimal growing conditions, equip your greenhouse with ventilation systems, fans, and heaters. Automatic vent openers and shade cloths can help regulate temperature and prevent overheating during hot summer days. In colder months, supplemental heating may be necessary to maintain the desired temperature. Electric heaters, propane heaters, or even passive solar heating methods can be used, depending on your setup and budget.

## Fall and Winter Gardening

Fall and winter gardening can be incredibly rewarding, providing fresh produce during months when most gardens lie dormant. With the right planning and techniques, you can enjoy a continuous harvest of hardy vegetables and greens.

## Planning Your Fall Garden

Start planning your fall garden in midsummer. Choose crops that can withstand colder temperatures and mature quickly. Cool-season vegetables such as kale, spinach, carrots, beets, and broccoli are excellent choices. Many of these crops actually develop better flavor when exposed to light frosts.

To ensure a continuous harvest, stagger your plantings by sowing seeds every few weeks. This technique, known as succession planting, allows you to harvest crops at different times rather than all at once.

Prepare your soil by adding compost or well-rotted manure to replenish nutrients depleted by summer crops. Remove any remaining summer plants and weeds to reduce competition for resources.

## Planting and Care

Direct sow seeds for root crops like carrots, beets, and radishes directly into the garden. For leafy greens and brassicas, start seeds indoors or in a cold frame, and transplant seedlings into the garden when they are large enough to handle.

As the weather cools, mulching becomes essential. Apply a thick layer of straw, leaves, or grass clippings around plants to insulate the soil and retain moisture. Mulch also helps to suppress weeds, which can still compete with your crops even in cooler weather.

Water your fall garden consistently, but be mindful not to overwater. Cooler temperatures reduce evaporation, so the soil retains moisture longer. Water early in the day to allow foliage to dry before nightfall, reducing the risk of fungal diseases.

## Protecting Crops

Use row covers or low tunnels made from hoops and clear plastic to protect plants from frost and wind. These structures create a microclimate that can be several degrees warmer than the outside air, extending the growing season by several weeks.

For even more protection, consider using double layers of row covers or combining row covers with cold frames. This method can protect crops from more severe frosts and extend the harvest well into winter.

Root crops such as carrots, parsnips, and beets can be left in the ground and harvested as needed. Cover them with a thick layer of mulch to prevent the ground from freezing and protect the roots from frost.

## Season Extension Techniques

In addition to cold frames, greenhouses, and row covers, several other season extension techniques can help you maximize your garden's yield and enjoy fresh produce year-round.

## Mulching

Mulching is one of the simplest and most effective ways to extend the growing season. A thick layer of mulch insulates the soil, retaining warmth and moisture. This is particularly beneficial for root crops, which can continue to grow and be harvested late into the season.

Straw, leaves, and grass clippings are excellent mulching materials. Apply mulch around the base of plants, ensuring a layer at least 2-3 inches thick. In colder climates, increase the thickness to 6-8 inches for better insulation.

## Raised Beds

Raised beds warm up faster in the spring and retain heat longer in the fall, extending the growing season. They also provide better drainage and soil structure, which can enhance plant growth.

Construct raised beds using untreated wood, bricks, or concrete blocks. Fill them with high-quality garden soil mixed with compost or well-rotted manure. The elevated height makes it easier to cover plants with row covers or cold frames for additional protection.

## Hoop Houses and Low Tunnels

Hoop houses and low tunnels are simple structures made from PVC or metal hoops covered with clear plastic or row cover fabric. They create a microclimate that is several degrees warmer than the surrounding air, protecting plants from frost and extending the growing season.

To construct a hoop house, drive metal or PVC pipes into the ground at regular intervals along the garden bed. Bend the pipes to form arches and cover them with clear plastic or row cover fabric. Secure the cover with clips or weights to prevent it from blowing away.

Low tunnels are similar but smaller and lower to the ground. They are ideal for protecting individual rows or smaller sections of the garden. Use the same construction method, but with shorter hoops and narrower covers.

## Hotbeds

Hotbeds use the heat generated from decomposing organic matter to warm the soil, creating an ideal environment for starting seeds or growing early crops. They are particularly useful in colder climates or for gardeners looking to get a head start on the growing season.

To create a hotbed, dig a trench about 2-3 feet deep and fill it with a layer of fresh manure and straw. Cover this layer with a few inches of garden soil. As the organic matter decomposes, it generates heat, warming the soil above.

Place a cold frame or hoop house over the hotbed to retain heat and protect plants from the elements. This method can extend the growing season by several weeks and provide a warm environment for starting seeds early in the spring.

## Using Thermal Mass

Thermal mass refers to materials that absorb, store, and release heat slowly. By incorporating thermal mass into your garden, you can create a more stable environment for your plants, extending the growing season.

Common thermal mass materials include water, stone, and concrete. Place large containers of water, such as barrels or tanks, in your greenhouse or cold frame. During the day, the water absorbs heat from the sun, and at night, it releases the stored heat, helping to maintain a stable temperature.

Stone or concrete paths and walls can also act as thermal mass, absorbing heat during the day and releasing it at night. These materials can help moderate temperature fluctuations and protect plants from frost.

## Companion Planting

Companion planting involves growing certain plants together to benefit each other. Some plants can provide shade, wind protection, or even trap heat for their companions, extending the growing season.

For example, tall plants like corn or sunflowers can provide shade and wind protection for smaller, more delicate plants. Ground covers like clover or creeping thyme can help retain soil moisture and warmth.

Incorporating companion planting into your garden design can create a more resilient and productive ecosystem, helping to maximize your yield and extend the growing season.

Maximizing your yield through season extension techniques is a vital part of successful homesteading. By understanding and implementing strategies such as cold frames, greenhouses, fall and winter gardening, and various other techniques, you can significantly increase your garden's productivity. These methods not only allow you to enjoy fresh produce year-round but also make your homestead more self-sufficient and resilient.

With careful planning, regular maintenance, and a willingness to experiment, you can push the boundaries of your growing season and make the most of your garden's potential. Embrace the challenge of extending your growing season and watch your homestead thrive in every season.

# Chapter 16

# Market Gardening and Selling Produce

Turning your garden into a profitable venture through market gardening and selling produce can be an incredibly rewarding experience. Market gardening focuses on growing a diverse range of crops intensively on a small scale, aiming for high yield and quality. This chapter will guide you through planning for market, the best practices for harvesting and post-harvest handling, and effective strategies for selling at farmers markets and through Community Supported Agriculture (CSA) programs.

## Planning for Market

Planning is a critical component of successful market gardening. It involves selecting crops, designing efficient garden layouts, managing planting schedules, and anticipating market demand.

### Selecting Crops

Choosing the right crops is crucial for market gardening. Consider the following factors when selecting crops for your garden:

**Market Demand:** Research local markets to determine which fruits and vegetables are in high demand. Visit farmers markets, talk to potential customers, and network with other farmers to get a sense of what sells well.

**Growing Conditions:** Select crops that are well-suited to your local climate and soil conditions. Crops that thrive in your environment will be more productive and require fewer resources.

**Seasonality:** Plan for a diverse range of crops that can be harvested throughout the growing season. This ensures a steady supply of produce for the market.

**Profitability:** Consider the potential profit margin of each crop. High-value crops like heirloom tomatoes, specialty greens, and herbs can be more profitable than lower-value staples.

**Variety:** Offering a variety of crops can attract more customers and provide a buffer against market fluctuations. Include a mix of staple vegetables, unique varieties, and value-added products.

## Designing Garden Layouts

An efficient garden layout maximizes productivity and simplifies management. Here are some key principles to consider when designing your market garden layout:

**Crop Rotation:** Implement a crop rotation plan to prevent soil depletion, reduce pest and disease pressure, and improve soil health. Rotate crops based on their botanical families and nutrient requirements.

**Intercropping:** Grow compatible crops together to maximize space and increase biodiversity. For example, plant quick-growing crops like radishes between slower-growing crops like tomatoes.

**Succession Planting:** Stagger plantings of the same crop at intervals to ensure a continuous harvest. This is especially important for crops with short harvest windows like lettuce and radishes.

**Raised Beds:** Use raised beds to improve soil drainage, reduce compaction, and increase soil temperature. Raised beds also make planting, weeding, and harvesting more efficient.

**Paths and Access:** Design wide, permanent paths to provide easy access to all parts of the garden. This facilitates efficient movement of tools, equipment, and harvested produce.

**Irrigation:** Plan an efficient irrigation system, such as drip irrigation, to provide consistent moisture to crops while conserving water.

## Managing Planting Schedules

Creating a detailed planting schedule is essential for timely planting, harvesting, and market supply. Follow these steps to develop a comprehensive planting schedule:

**Frost Dates:** Determine your local average last frost date in spring and first frost date in fall. Use this information to calculate planting and harvesting dates for each crop.

**Seed Starting:** Start seeds indoors or in a greenhouse for crops that require a longer growing season. Transplant seedlings to the garden after the danger of frost has passed.

**Direct Sowing:** Direct sow seeds for crops that do not transplant well, such as root vegetables and some greens. Follow the recommended planting depth and spacing for each crop.

**Successive Plantings:** Plan for successive plantings of fast-maturing crops. For example, sow a new batch of lettuce every two weeks to ensure a continuous supply.

**Record Keeping:** Maintain detailed records of planting dates, germination rates, and harvest dates. Use this information to refine your planting schedule each year.

## Harvesting and Post-Harvest Handling

Proper harvesting and post-harvest handling are essential to maintain the quality and freshness of your produce. Follow best practices to ensure your crops reach the market in peak condition.

## Harvesting Techniques

Different crops require specific harvesting techniques to ensure they are picked at the right time and handled gently to avoid damage.

**Leafy Greens:** Harvest leafy greens like lettuce, spinach, and kale early in the morning when they are crisp and hydrated. Use a sharp knife or scissors to cut leaves, leaving the base of the plant intact for regrowth.

**Root Vegetables:** Harvest root vegetables like carrots, beets, and radishes when they reach the desired size. Loosen the soil around the roots with a garden fork before gently pulling them out.

**Fruiting Vegetables:** Pick fruiting vegetables like tomatoes, peppers, and cucumbers when they are fully ripe but still firm. Use pruning shears or twist the fruit gently to avoid damaging the plant.

**Herbs:** Harvest herbs like basil, parsley, and cilantro regularly to encourage new growth. Cut stems just above a leaf node to promote bushier plants.

**Perishable Crops:** Harvest perishable crops like berries and peas frequently to ensure peak freshness. Handle these crops gently to avoid bruising.

## Post-Harvest Handling

Proper post-harvest handling is crucial to maintain the quality and shelf life of your produce. Follow these steps to ensure your crops are market-ready:

**Cooling:** Cool harvested produce as quickly as possible to remove field heat and slow the respiration rate. Use methods like hydrocooling (immersing in cold water) or air cooling (placing in a cool, shaded area).

**Washing:** Wash produce thoroughly to remove dirt and debris. Use clean, cold water and handle crops gently to avoid damage. Some crops, like leafy greens, may benefit from a final rinse in cold, chlorinated water to reduce microbial load.

**Drying:** Dry produce thoroughly after washing to prevent mold and spoilage. Use clean towels, salad spinners, or mesh racks for air drying.

**Grading and Sorting:** Sort produce by size, quality, and ripeness. Remove any damaged or blemished items. Uniformity in appearance can increase the appeal of your produce at the market.

**Packaging:** Use appropriate packaging to protect produce during transport and display. Options include plastic bags, clamshell containers, and waxed boxes. Ensure packaging is clean and food-safe.

**Storage:** Store harvested produce at the appropriate temperature and humidity levels. Use coolers, refrigerators, or root cellars to extend shelf life. Monitor storage conditions regularly to prevent spoilage.

## Record Keeping

Maintain detailed records of harvest dates, quantities, and post-harvest handling practices. This information can help you track the productivity of your crops, identify areas for improvement, and provide transparency to your customers.

## Selling at Farmers Markets and CSA

Once your produce is harvested and prepared for sale, the next step is marketing and selling it effectively. Farmers markets and Community Supported Agriculture (CSA) programs are popular avenues for selling fresh, local produce.

## Selling at Farmers Markets

Farmers markets offer a direct-to-consumer sales platform, allowing you to interact with customers, build relationships, and receive immediate feedback.

## Preparing for Market

Preparation is key to a successful farmers market experience. Here are some steps to ensure you are ready:

**Market Research:** Visit local farmers markets to understand the market dynamics, customer preferences, and competitor offerings. Choose markets that align with your products and target audience.

**Market Rules:** Familiarize yourself with the market's rules and regulations, including stall fees, setup times, and product standards. Ensure you comply with all requirements, including food safety regulations and permits.

**Product Selection:** Offer a diverse range of high-quality produce to attract customers. Highlight unique or specialty items that differentiate you from other vendors.

**Pricing:** Set competitive prices based on market research, production costs, and perceived value. Consider offering bundle deals or discounts for bulk purchases.

**Display:** Invest in attractive and functional display equipment, such as tables, tents, and signage. Arrange produce neatly, with the most eye-catching items prominently displayed. Use baskets, crates, and tiers to create an appealing visual presentation.

## Engaging with Customers

Building relationships with customers is essential for repeat business and word-of-mouth referrals. Here are some tips for effective customer engagement:

**Personal Interaction:** Greet customers warmly, make eye contact, and engage in friendly conversation. Share your story and passion for farming to create a personal connection.

**Product Knowledge:** Be knowledgeable about your products and their uses. Provide cooking tips, recipes, and storage advice to help customers make the most of their purchases.

**Sampling:** Offer samples of your produce to allow customers to taste the quality and flavor. Ensure samples are presented hygienically and comply with market regulations.

**Education:** Educate customers about the benefits of buying local, seasonal produce. Highlight the nutritional value, freshness, and environmental benefits of your products.

**Feedback:** Encourage customer feedback and be responsive to their needs and preferences. Use feedback to improve your products and services.

## Marketing Strategies

Effective marketing can boost your visibility and sales at farmers markets. Consider the following strategies:

**Branding:** Develop a strong brand identity, including a memorable farm name, logo, and tagline. Use consistent branding across all marketing materials, including signage, business cards, and social media.

**Social Media:** Use social media platforms like Facebook, Instagram, and Twitter to promote your market appearances, share behind-the-scenes content, and engage with customers. Post regular updates about your products, market locations, and special promotions.

**Loyalty Programs:** Implement loyalty programs to reward repeat customers. Offer discounts, free items, or exclusive access to new products for loyal patrons.

**Collaborations:** Partner with other vendors or local businesses to cross-promote products and services. Joint promotions, bundled products, and co-hosted events can attract new customers and increase sales.

## Community Supported Agriculture (CSA)

Community Supported Agriculture (CSA) programs provide a direct connection between farmers and consumers. Customers purchase a share of the farm's harvest in advance and receive regular deliveries of fresh produce throughout the growing season.

### Setting Up a CSA Program

Setting up a successful CSA program requires careful planning and clear communication with members. Here are the key steps:

**Planning and Preparation:** Determine the size and scope of your CSA program based on your farm's production capacity and market demand. Decide on the number of shares you will offer, the length of the season, and the frequency of deliveries.

**Membership Structure:** Define the terms of your CSA membership, including the cost of shares, payment schedules, and delivery or pickup options. Offer flexible payment plans to accommodate different budgets.

**Crop Planning:** Plan your crop production to ensure a diverse and consistent supply of produce throughout the season. Consider the preferences of your members and include a mix of staple vegetables, unique varieties, and occasional surprises.

**Marketing and Recruitment:** Promote your CSA program through your website, social media, farmers markets, and local community organizations. Highlight the benefits of CSA membership, such as fresh, seasonal produce, support for local agriculture, and a direct connection to the farm.

## Managing a CSA Program

Effective management is essential to maintaining member satisfaction and ensuring the smooth operation of your CSA program. Here are some best practices:

**Member Communication:** Maintain regular communication with your members through newsletters, emails, and social media. Provide updates on crop progress, upcoming deliveries, and farm events. Share recipes, storage tips, and cooking ideas to help members make the most of their shares.

**Packing and Distribution:** Pack CSA shares carefully to ensure produce is fresh and undamaged. Use sturdy, reusable containers and include a list of the week's items. Coordinate distribution logistics, whether through home delivery, central pickup locations, or on-farm pickups.

**Member Engagement:** Foster a sense of community among your members by hosting farm tours, potlucks, and educational workshops. Encourage members to participate in farm activities and provide feedback on their CSA experience.

**Quality Control:** Monitor the quality of the produce included in CSA shares and address any issues promptly. Ensure members receive a fair share of high-quality produce, even during challenging growing conditions.

**Record Keeping:** Keep detailed records of crop production, harvest quantities, and member feedback. Use this data to refine your crop planning, improve member satisfaction, and enhance the overall efficiency of your CSA program.

Market gardening and selling produce can be a fulfilling and profitable venture for homesteaders. By carefully planning your market garden, implementing best practices for harvesting and post-harvest handling, and effectively marketing your produce through farmers markets and CSA programs, you can create a sustainable and thriving business. With dedication, attention to detail, and a commitment to quality, you can maximize your yield, build strong relationships with customers, and enjoy the rewards of growing and selling fresh, local produce.

# Chapter 17

# **Perennial Plants and Trees**

Perennial plants and trees are foundational to a sustainable and productive homestead. Unlike annuals that need replanting each year, perennials come back season after season, providing a steady supply of food, enhancing biodiversity, and reducing the need for extensive soil preparation. This chapter delves into the benefits of perennials, the essentials of planting and caring for fruit trees and berry bushes, and strategies for integrating perennials into your homestead.

## **Benefits of Perennials**

Perennial plants offer numerous advantages that make them indispensable to any homestead. They are resilient, require less maintenance than annuals, and contribute significantly to the ecosystem's health.

## **Reduced Labor and Maintenance**

One of the primary benefits of perennials is the reduction in labor and maintenance. Once established, perennials require minimal effort compared to annuals. They do not need to be replanted every year, which saves time and labor. Perennials also tend to have deeper root systems that help them access water and nutrients more efficiently, reducing the need for frequent watering and fertilizing.

## Soil Health and Erosion Control

Perennials play a crucial role in maintaining soil health. Their extensive root systems help improve soil structure, enhance aeration, and increase organic matter content. These roots also help prevent soil erosion by stabilizing the soil and reducing runoff. Over time, perennials contribute to a healthier, more resilient soil ecosystem.

## Biodiversity and Wildlife Habitat

Perennials support biodiversity by providing habitat and food for a variety of wildlife. Flowering perennials attract pollinators like bees, butterflies, and hummingbirds, which are essential for the pollination of many crops. Perennial plants also offer shelter and food for beneficial insects, birds, and other wildlife, creating a balanced and thriving ecosystem on your homestead.

## Sustainable Food Production

Perennial crops can provide a reliable source of food year after year. Fruit trees, berry bushes, and perennial vegetables can yield abundant harvests with proper care. This consistent food production reduces dependence on annual crops and contributes to food security on your homestead.

# Planting and Caring for Fruit Trees and Berry Bushes

Fruit trees and berry bushes are among the most rewarding perennials to grow. They require careful planning, proper planting, and ongoing care to thrive and produce abundant harvests.

## Selecting Fruit Trees and Berry Bushes

Choosing the right varieties of fruit trees and berry bushes is crucial for success. Consider factors such as climate, soil type, space, and intended use of the fruit when selecting varieties.

**Climate Suitability**: Select varieties that are well-suited to your local climate. Consider factors such as chill hours (the number of hours below 45°F needed to break dormancy), frost tolerance, and heat resistance. Local nurseries and extension services can provide valuable guidance on suitable varieties for your area.

**Soil Type and Drainage**: Ensure that the selected varieties are compatible with your soil type. Fruit trees and berry bushes generally prefer well-drained soil. Conduct a soil test to determine soil pH and nutrient levels, and amend the soil as needed to create optimal growing conditions.

**Space and Growth Habit**: Consider the mature size and growth habit of the trees and bushes. Ensure that you have enough space to accommodate their full size, and plan for proper spacing to allow for adequate air circulation and sunlight penetration.

**Pollination Requirements**: Some fruit trees require cross-pollination to produce fruit. Plant at least two compatible varieties within close proximity to ensure successful pollination. Berry bushes typically do not require cross-pollination, but planting multiple varieties can improve yields.

## Planting Fruit Trees

Proper planting techniques are essential for the successful establishment and growth of fruit trees. Follow these steps to plant fruit trees effectively:

**Site Preparation**: Choose a sunny location with well-drained soil. Clear the area of weeds and debris. If soil drainage is poor, consider creating raised beds or mounds to improve drainage.

**Digging the Hole**: Dig a hole that is twice as wide and just as deep as the tree's root ball. This allows the roots to spread easily and establish quickly. Loosen the soil at the bottom and sides of the hole to encourage root growth.

**Planting the Tree**: Place the tree in the hole, ensuring that the graft union (the swollen area where the tree was grafted onto the rootstock) is above the soil line. Backfill the hole with the excavated soil, gently firming it around the roots to eliminate air pockets.

**Watering**: Water the tree thoroughly after planting to settle the soil and ensure good root-to-soil contact. Continue to water regularly, especially during the first year, to establish a strong root system.

**Mulching**: Apply a layer of organic mulch, such as straw, wood chips, or compost, around the base of the tree. Mulching helps retain soil moisture, suppress weeds, and regulate soil temperature. Keep the mulch a few inches away from the trunk to prevent rot.

**Staking**: Stake the tree if necessary to provide support and protect it from wind damage. Use soft ties to secure the tree to the stake, allowing some flexibility for natural movement.

## Caring for Fruit Trees

Consistent care is essential for the health and productivity of fruit trees. Key aspects of fruit tree care include watering, fertilizing, pruning, and pest management.

**Watering**: Fruit trees need consistent moisture, especially during dry periods. Water deeply and infrequently to encourage deep root growth. Young trees require more frequent watering, while established trees can tolerate longer intervals between waterings.

**Fertilizing**: Provide balanced nutrition to support healthy growth and fruit production. Conduct a soil test to determine nutrient deficiencies and apply fertilizers accordingly. Use organic fertilizers, such as compost or well-rotted manure, to improve soil fertility.

**Pruning**: Prune fruit trees annually to maintain their shape, improve air circulation, and remove dead or diseased branches. Pruning also helps manage tree size and encourages the production of high-quality fruit. Learn the specific pruning techniques for each type of fruit tree to ensure optimal results.

**Pest and Disease Management**: Monitor fruit trees regularly for signs of pests and diseases. Implement integrated pest management (IPM) practices, such as encouraging beneficial insects, using organic sprays, and removing infected plant material. Proper sanitation and cultural practices, such as pruning and mulching, can also help prevent pest and disease problems.

## Planting Berry Bushes

Berry bushes, such as blueberries, raspberries, and blackberries, are excellent additions to a homestead. They are relatively easy to grow and can produce abundant harvests with proper care.

**Site Selection**: Choose a sunny location with well-drained soil for planting berry bushes. Blueberries prefer acidic soil with a pH between 4.5 and 5.5, while raspberries and blackberries thrive in slightly acidic to neutral soil (pH 5.5 to 7.0).

**Soil Preparation**: Prepare the soil by incorporating organic matter, such as compost or peat moss, to improve soil structure and fertility. For blueberries, amend the soil with elemental sulfur or pine needles to lower the pH if necessary.

**Planting Depth**: Plant berry bushes at the same depth they were growing in their containers or nurseries. Space blueberry bushes 4 to 6 feet apart and raspberries and blackberries 2 to 4 feet apart. Allow sufficient space between rows for easy access and air circulation.

**Watering and Mulching**: Water berry bushes thoroughly after planting and continue to water regularly, especially during dry periods. Apply a layer of mulch around the base of the plants to retain soil moisture, suppress weeds, and regulate soil temperature.

## Caring for Berry Bushes

Proper care ensures healthy growth and high yields from berry bushes. Key aspects of berry bush care include watering, fertilizing, pruning, and pest management.

**Watering**: Berry bushes require consistent moisture, particularly during fruit development. Water deeply and regularly, aiming for about 1 inch of water per week. Mulching helps retain soil moisture and reduces the need for frequent watering.

**Fertilizing**: Fertilize berry bushes annually to support healthy growth and fruit production. Use a balanced fertilizer formulated for berries or apply organic fertilizers, such as compost or well-rotted manure. Blueberries benefit from acidifying fertilizers, such as those containing ammonium sulfate.

**Pruning**: Prune berry bushes annually to maintain their shape, remove dead or diseased canes, and encourage new growth. Pruning techniques vary by type of berry bush. For example, prune blueberry bushes to remove older, less productive canes and encourage new shoots. For raspberries and blackberries, remove old canes that have fruited and thin out new canes to improve air circulation.

**Pest and Disease Management**: Monitor berry bushes regularly for signs of pests and diseases. Implement IPM practices, such as encouraging beneficial insects, using organic sprays, and removing infected plant material. Proper sanitation and cultural practices, such as pruning and mulching, can help prevent pest and disease problems.

## Integrating Perennials into Your Homestead

Integrating perennials into your homestead can enhance its productivity, resilience, and ecological balance. By strategically incorporating fruit trees, berry bushes, and other perennials, you can create a diverse and sustainable landscape.

# Designing a Perennial Garden

Designing a perennial garden involves thoughtful planning to ensure that plants thrive and complement each other. Consider the following elements when designing your perennial garden:

**Site Analysis**: Assess the site for sunlight, soil type, drainage, and microclimates. Understanding the site's characteristics helps you select suitable plants and determine the best layout.

**Plant Selection**: Choose a mix of fruit trees, berry bushes, and other perennials that are well-suited to your climate and soil. Consider the mature size, growth habit, and compatibility of the plants.

**Layering**: Design your garden using the concept of layering, which mimics natural ecosystems. Layering involves planting different types of plants at various heights to maximize space and resources. For example, plant tall fruit trees as the canopy layer, berry bushes and shrubs as the understory layer, and herbs and groundcovers as the ground layer.

**Companion Planting**: Integrate companion plants that benefit each other by improving soil health, deterring pests, or enhancing pollination. For example, plant garlic and chives around fruit trees to repel pests, or interplant strawberries with asparagus to maximize space and reduce weed competition.

**Water Management**: Plan for efficient water management by grouping plants with similar water needs together. Implement irrigation systems, such as drip irrigation or soaker hoses, to provide consistent moisture while conserving water.

**Mulching and Soil Improvement**: Apply mulch to retain soil moisture, suppress weeds, and improve soil fertility. Use organic matter, such as compost and cover crops, to enhance soil structure and nutrient levels.

## Incorporating Edible Landscapes

Edible landscapes combine aesthetic beauty with food production, creating a multifunctional and attractive garden. Here are some ways to incorporate edible plants into your landscape:

**Orchards**: Establish a fruit orchard by planting a variety of fruit trees in a designated area. Space the trees appropriately and consider interplanting with berry bushes or groundcovers to maximize productivity and biodiversity.

**Edible Hedges**: Use berry bushes or dwarf fruit trees to create edible hedges that provide privacy, wind protection, and food. For example, plant blueberry bushes or dwarf apple trees along property lines or as garden borders.

**Perennial Borders**: Design perennial borders with a mix of fruiting shrubs, herbs, and flowers. This not only adds visual interest but also provides habitat for pollinators and beneficial insects. Consider planting lavender, rosemary, and sage alongside berry bushes for a beautiful and functional border.

**Kitchen Gardens**: Create a kitchen garden near your home with a mix of perennial herbs, vegetables, and fruits. Incorporate raised beds, containers, and vertical structures to optimize space and accessibility. Planting herbs like thyme, oregano, and mint alongside perennial vegetables like rhubarb and asparagus can create a productive and attractive kitchen garden.

**Food Forests**: Design a food forest by mimicking the structure and diversity of a natural forest. Include multiple layers of plants, from tall fruit trees to groundcovers, to create a self-sustaining ecosystem. A food forest can provide a continuous supply of food, enhance biodiversity, and improve soil health.

## Enhancing Biodiversity and Ecosystem Health

Integrating perennials into your homestead can significantly enhance biodiversity and ecosystem health. Here are some strategies to promote a healthy and balanced ecosystem:

**Pollinator Gardens**: Create pollinator gardens with a variety of flowering perennials that provide nectar and pollen throughout the growing season. Include plants like bee balm, coneflower, and milkweed to attract bees, butterflies, and other pollinators.

**Beneficial Insects**: Encourage beneficial insects by planting insectary plants, such as dill, fennel, and yarrow, which provide habitat and food for predatory insects. These beneficial insects help control pest populations naturally.

**Habitat Creation**: Provide habitat for wildlife by incorporating a mix of plants that offer food, shelter, and nesting sites. Include native plants, berry bushes, and fruit trees that support birds, small mammals, and insects.

**Water Features**: Add water features, such as ponds or birdbaths, to attract wildlife and enhance biodiversity. Water features provide drinking and bathing sources for birds and insects and create a balanced ecosystem.

**Soil Health**: Maintain soil health by practicing crop rotation, cover cropping, and mulching. Healthy soil supports diverse plant and microbial communities, which in turn enhance overall ecosystem health.

## Sustainable Practices

Implementing sustainable practices on your homestead ensures long-term productivity and environmental stewardship. Here are some key sustainable practices for integrating perennials:

**Organic Farming**: Avoid synthetic pesticides and fertilizers, and use organic methods to manage pests and improve soil fertility. Organic farming promotes soil health, biodiversity, and ecosystem balance.

**Water Conservation**: Implement water conservation techniques, such as rainwater harvesting, drip irrigation, and mulching, to reduce water use and ensure sustainable water management.

**Permaculture Principles**: Apply permaculture principles to design a self-sustaining and resilient homestead. Principles like observing and interacting with nature, using renewable resources, and creating closed-loop systems can enhance sustainability.

**Composting**: Compost kitchen scraps, garden waste, and animal manure to create nutrient-rich compost for your garden. Composting reduces waste, improves soil fertility, and supports healthy plant growth.

**Biodiversity**: Promote biodiversity by planting a wide variety of perennials and incorporating native plants. Biodiversity enhances ecosystem resilience, reduces pest and disease pressure, and improves overall garden health.

## Conclusion

Perennial plants and trees are essential components of a sustainable and productive homestead. Their benefits include reduced labor and maintenance, improved soil health, enhanced biodiversity, and reliable food production. By carefully selecting, planting, and caring for fruit trees and berry bushes, you can enjoy abundant harvests year after year. Integrating perennials into your homestead through thoughtful design and sustainable practices creates a resilient and thriving landscape that supports both your needs and the environment. Embrace the power of perennials to transform your homestead into a productive and sustainable haven.

# Chapter 18
# Advanced Techniques and Future Planning

Advanced techniques in soil fertility and future planning are essential for taking your homesteading to the next level. By understanding and implementing practices like cover cropping, using green manures, and incorporating biochar and other soil amendments, you can create a more sustainable and productive homestead. These methods not only enhance soil health but also improve crop yields, reduce pest and disease pressures, and contribute to the overall resilience of your farming system.

## Advanced Soil Fertility Practices

Soil fertility is the cornerstone of successful gardening and farming. Advanced soil fertility practices involve understanding the intricate dynamics of soil ecosystems and using techniques that enhance soil health and productivity over the long term.

## Understanding Soil Fertility

Soil fertility refers to the soil's ability to provide essential nutrients to plants. Healthy, fertile soil is rich in organic matter, has a balanced pH, and contains a diverse array of microorganisms that aid in nutrient cycling and disease suppression.

**Soil Composition**: Soil is composed of mineral particles (sand, silt, and clay), organic matter, water, and air. The proportion of these components affects the soil's texture, structure, and ability to retain water and nutrients.

**Soil pH**: Soil pH affects the availability of nutrients to plants. Most crops prefer a pH range of 6.0 to 7.0. Regular soil testing is crucial to monitor pH and make adjustments as necessary.

**Nutrient Cycling**: Soil microorganisms play a vital role in breaking down organic matter and releasing nutrients in forms that plants can absorb. Healthy soil has a thriving community of bacteria, fungi, and other microorganisms.

## Enhancing Soil Fertility

To enhance soil fertility, it's essential to focus on building organic matter, maintaining a balanced pH, and fostering a diverse soil ecosystem.

**Adding Organic Matter**: Organic matter improves soil structure, water retention, and nutrient availability. Compost, well-rotted manure, and cover crops are excellent sources of organic matter.

**Balancing pH**: Lime can be added to raise soil pH, while sulfur can lower it. Regular soil testing helps determine the need for these amendments.

**Promoting Microbial Activity**: Avoid practices that harm soil microorganisms, such as excessive tillage and the use of synthetic chemicals.

Instead, use organic fertilizers and soil conditioners to support a healthy microbial community.

## Cover Cropping

Cover cropping is the practice of growing crops specifically to improve soil health rather than for harvest. Cover crops are planted during periods when the main crops are not growing, such as in the winter or between crop rotations.

### Benefits of Cover Cropping

Cover crops offer numerous benefits, including improving soil structure, preventing erosion, suppressing weeds, and enhancing nutrient cycling.

**Soil Structure**: Cover crops, especially those with deep root systems, help break up compacted soil and improve its structure. This enhances water infiltration and root penetration for subsequent crops.

**Erosion Control**: Cover crops protect the soil surface from erosion caused by wind and water. Their roots help bind the soil, reducing runoff and sediment loss.

**Weed Suppression**: Dense cover crop stands shade the soil, outcompeting weeds for light and nutrients. Some cover crops, like rye and buckwheat, also release allelopathic compounds that inhibit weed germination.

**Nutrient Cycling**: Cover crops capture and recycle nutrients that might otherwise be lost to leaching. Leguminous cover crops, such as clover and vetch, fix atmospheric nitrogen, enriching the soil with this essential nutrient.

## Choosing Cover Crops

The choice of cover crops depends on your specific goals, climate, and soil conditions. Here are some common cover crops and their benefits:

**Legumes**: Leguminous cover crops, such as clover, vetch, and peas, fix nitrogen in the soil, reducing the need for synthetic fertilizers. They also improve soil structure and add organic matter.

**Grasses**: Grasses like rye, oats, and barley are excellent for erosion control and weed suppression. Their fibrous root systems improve soil structure and organic matter content.

**Brassicas**: Brassicas, such as radishes and mustard, are effective at breaking up compacted soil and suppressing pests and diseases. They also add organic matter and improve nutrient cycling.

**Mixes**: Combining different cover crops can maximize their benefits. For example, a mix of legumes and grasses can provide nitrogen fixation, erosion control, and organic matter addition.

## Planting and Managing Cover Crops

Planting and managing cover crops involve selecting the appropriate species, timing the planting, and incorporating the cover crop into the soil.

**Timing**: Plant cover crops in the fall after harvesting the main crops or in early spring before planting. Timing is crucial to ensure that cover crops establish well and provide maximum benefits.

**Seeding Rates**: Follow recommended seeding rates for each cover crop species to achieve optimal stand density. Over-seeding can lead to competition among plants, while under-seeding may result in poor coverage.

**Termination**: Terminate cover crops before they set seed to prevent them from becoming weeds. Methods for termination include mowing, rolling, and incorporating the cover crop into the soil using tillage or no-till techniques.

## Green Manures

Green manures are crops grown specifically to be incorporated into the soil to improve its fertility and structure. They are usually plowed under while still green and lush, providing a quick source of organic matter and nutrients.

### Benefits of Green Manures

Green manures offer several advantages for soil health, including enhancing organic matter, improving nutrient availability, and suppressing weeds and pests.

**Organic Matter Addition**: Green manures decompose quickly, adding fresh organic matter to the soil. This improves soil structure, water retention, and microbial activity.

**Nutrient Availability**: As green manures decompose, they release nutrients that are readily available to subsequent crops. This can reduce the need for synthetic fertilizers and improve soil fertility.

**Weed and Pest Suppression**: Some green manure crops, like mustard and radish, have biofumigant properties that suppress soil-borne pests and diseases. Additionally, their dense growth can shade out weeds.

## Selecting Green Manure Crops

Choosing the right green manure crops depends on your specific needs and growing conditions. Here are some common green manure crops and their benefits:

**Legumes**: Leguminous green manures, such as alfalfa, clover, and soybeans, fix nitrogen and improve soil fertility. They also add organic matter and enhance soil structure.

**Grasses**: Grasses like rye, oats, and wheat are excellent for adding organic matter and improving soil structure. They are also effective at scavenging residual nutrients and preventing erosion.

**Brassicas**: Brassicas, such as mustard and radish, are effective at breaking up compacted soil and suppressing pests and diseases. They add organic matter and improve nutrient cycling.

## Planting and Managing Green Manures

The successful use of green manures involves proper planting, management, and incorporation into the soil.

**Timing**: Plant green manures during periods when the soil would otherwise be bare, such as in the fall after harvest or in early spring before planting. Timing is crucial to ensure that green manures establish well and provide maximum benefits.

**Seeding Rates**: Follow recommended seeding rates for each green manure crop to achieve optimal stand density. Over-seeding can lead to competition among plants, while under-seeding may result in poor coverage.

**Termination and Incorporation**: Terminate green manures before they set seed to prevent them from becoming weeds. Incorporate the green manure into the soil while it is still green and lush, using tillage or no-till techniques. This allows for quick decomposition and nutrient release.

# Biochar and Other Soil Amendments

Biochar and other soil amendments can significantly improve soil health, enhance nutrient availability, and increase crop productivity. Understanding how to use these amendments effectively can lead to more sustainable and productive farming practices.

## Biochar

Biochar is a stable form of carbon produced by heating organic material (such as wood or crop residues) in the absence of oxygen. It is used as a soil amendment to improve soil fertility and sequester carbon.

## Benefits of Biochar

Biochar offers several benefits for soil health and crop productivity:

**Soil Structure**: Biochar improves soil structure by increasing porosity and enhancing water and nutrient retention. This is particularly beneficial in sandy or compacted soils.

**Nutrient Availability**: Biochar can enhance nutrient availability by retaining nutrients in the root zone and reducing leaching. It also provides a habitat for beneficial soil microorganisms that aid in nutrient cycling.

**Carbon Sequestration**: Biochar sequesters carbon in the soil for hundreds to thousands of years, helping to mitigate climate change by reducing atmospheric carbon dioxide levels.

**pH Regulation**: Biochar can help regulate soil pH, making it more neutral, which benefits most crops.

## Producing and Applying Biochar

Producing and applying biochar involves selecting feedstock, pyrolysis (the process of heating organic material in the absence of oxygen), and incorporating biochar into the soil.

**Feedstock Selection**: Choose feedstock that is readily available and free of contaminants. Common feedstocks include wood chips, crop residues, and manure.

**Pyrolysis Process**: Produce biochar through pyrolysis, which can be done using simple or advanced equipment. The key is to achieve high temperatures (300-700°C) in the absence of oxygen to convert the organic material into stable carbon.

**Application Rates**: Apply biochar at a rate of 5-20% by volume of the soil. Incorporate biochar into the soil by mixing it thoroughly to ensure even distribution.

**Activation**: Activate biochar by mixing it with compost or manure before application. This charges the biochar with nutrients and microorganisms, enhancing its effectiveness as a soil amendment.

## Other Soil Amendments

In addition to biochar, several other soil amendments can improve soil health and fertility. These amendments include compost, manure, rock dust, and organic fertilizers.

## Compost

Compost is decomposed organic matter that enriches the soil with nutrients and beneficial microorganisms. It improves soil structure, increases water retention, and enhances nutrient availability.

**Benefits**: Compost adds organic matter, improves soil fertility, and enhances microbial activity. It can also suppress soil-borne diseases and pests.

**Application**: Apply compost by spreading it over the soil surface or incorporating it into the topsoil. Use compost as a mulch, a soil conditioner, or a component of potting mixes.

## Manure

Manure is animal waste that provides essential nutrients and organic matter to the soil. It improves soil structure, fertility, and microbial activity.

**Types**: Different types of manure (e.g., cow, horse, chicken) have varying nutrient compositions. Choose manure based on availability and nutrient needs.

**Application**: Apply well-rotted manure to avoid burning plants with fresh manure. Incorporate manure into the soil before planting or use it as a top dressing during the growing season.

## Rock Dust

Rock dust, such as basalt or granite dust, is a mineral amendment that provides trace minerals and improves soil structure.

**Benefits**: Rock dust adds essential trace minerals, such as calcium, magnesium, and potassium, that may be deficient in the soil. It also improves soil structure and enhances microbial activity.

**Application**: Apply rock dust by spreading it over the soil surface and incorporating it into the topsoil. Use rock dust as a long-term soil conditioner, applying it every few years.

## Organic Fertilizers

Organic fertilizers are derived from natural sources, such as plant or animal materials, and provide essential nutrients to the soil.

1. **Types**: Common organic fertilizers include bone meal, blood meal, fish emulsion, and kelp meal. Each type has a different nutrient composition and release rate.

2. **Application**: Apply organic fertilizers based on soil test results and crop nutrient needs. Incorporate fertilizers into the soil before planting or use them as a side dressing during the growing season.

## Integrating Advanced Techniques into Your Homestead

Integrating advanced soil fertility practices, cover cropping, green manures, biochar, and other soil amendments into your homestead requires careful planning and execution. These practices can significantly enhance soil health, improve crop yields, and contribute to the sustainability and resilience of your farming system.

## Planning and Implementation

Developing a comprehensive plan for integrating advanced techniques involves assessing your current soil health, identifying goals, and implementing practices that align with those goals.

**Soil Testing**: Conduct regular soil tests to assess soil pH, nutrient levels, and organic matter content. Use this information to make informed decisions about soil amendments and fertility practices.

**Setting Goals**: Identify specific goals for soil health, crop productivity, and sustainability. Goals may include increasing organic matter, improving soil structure, enhancing nutrient availability, or reducing reliance on synthetic inputs.

**Selecting Practices**: Choose advanced techniques that align with your goals and are suitable for your soil type, climate, and crop rotations. Consider practices like cover cropping, green manures, biochar application, and organic fertilization.

## Monitoring and Adjusting

Continuous monitoring and adjustment are essential to ensure the success of advanced techniques. Regularly assess the effectiveness of practices and make necessary adjustments to optimize results.

**Record Keeping**: Maintain detailed records of soil test results, amendments applied, cover crop species used, and crop yields. Use this data to track progress and identify areas for improvement.

**Observations**: Observe changes in soil health, crop growth, and pest and disease pressures. Look for signs of improved soil structure, increased microbial activity, and enhanced crop vigor.

**Adjustments**: Adjust practices based on observations and data. For example, if soil pH remains imbalanced, consider additional amendments or changes in crop rotation. If cover crops are not providing desired benefits, try different species or planting times.

## Long-Term Sustainability

The ultimate goal of integrating advanced techniques is to achieve long-term sustainability and resilience on your homestead. This involves creating a farming system that maintains soil health, supports biodiversity, and produces high-quality crops year after year.

**Diversity**: Incorporate a diverse range of crops, cover crops, and soil amendments to enhance ecosystem resilience and reduce the risk of pests and diseases.

**Soil Stewardship**: Prioritize soil health in all farming decisions. Healthy soil is the foundation of a productive and sustainable homestead.

**Education and Adaptation**: Stay informed about new research and innovations in soil fertility and sustainable agriculture. Be willing to adapt and refine practices as needed to improve results.

Advanced soil fertility practices, cover cropping, green manures, biochar, and other soil amendments are powerful tools for enhancing the productivity and sustainability of your homestead. By understanding and implementing these techniques, you can create a thriving farming system that supports healthy crops, resilient ecosystems, and long-term soil health. Embrace these advanced practices to take your homesteading to the next level and ensure a productive and sustainable future for your land.

# Chapter 19

# Aquaponics and Hydroponics

Aquaponics and hydroponics are innovative farming techniques that can enhance your homestead's productivity and sustainability. These soil-less systems allow you to grow a wide range of crops in controlled environments, making them ideal for maximizing yield in limited spaces. In this chapter, we will explore the basics of aquaponics, the steps to set up a hydroponic system, and how to integrate aquaponics into your homestead.

## Basics of Aquaponics

Aquaponics is a sustainable farming method that combines aquaculture (raising fish) with hydroponics (growing plants in water). In an aquaponic system, fish waste provides nutrients for the plants, and the plants, in turn, help filter and clean the water for the fish. This symbiotic relationship creates a closed-loop system that is efficient and environmentally friendly.

### How Aquaponics Works

At its core, an aquaponics system consists of three main components: a fish tank, a grow bed, and a water pump to circulate water between them. Here is a detailed look at how these components interact:

1. **Fish Tank**: The fish tank houses the fish, which produce waste rich in ammonia. This waste is the primary nutrient source for the plants.

2. **Grow Bed**: The grow bed is where the plants are cultivated. It is filled with a growing medium, such as gravel or clay pebbles, that supports the plants and allows water to flow through it.

3. **Water Pump**: The water pump circulates water from the fish tank to the grow bed and back. As water flows through the grow bed, it is filtered by the plants and growing medium before returning to the fish tank.

The key to a successful aquaponics system is the nitrogen cycle, which involves the conversion of ammonia from fish waste into nitrites and then into nitrates by beneficial bacteria. These nitrates are readily absorbed by the plants, providing them with essential nutrients.

## Types of Aquaponic Systems

There are several types of aquaponic systems, each with its own advantages and applications. The three most common types are:

1. **Media-Based Aquaponics**: This system uses a growing medium, such as gravel or clay pebbles, to support the plants. Water from the fish tank is pumped into the grow bed, where it flows through the medium and is filtered by the plants before returning to the

fish tank. Media-based systems are simple to set up and maintain, making them ideal for beginners.

2. **Nutrient Film Technique (NFT)**: In an NFT system, plants are grown in channels with a thin film of water flowing over their roots. This water contains nutrients from the fish waste. NFT systems are efficient and well-suited for growing leafy greens and herbs but are more complex to set up and require precise water flow management.

3. **Deep Water Culture (DWC)**: Also known as raft systems, DWC involves floating plants on rafts in a deep tank filled with nutrient-rich water. The plant roots dangle into the water, absorbing nutrients directly. DWC systems are highly productive and ideal for large-scale production of leafy greens, but they require careful oxygenation of the water to prevent root rot.

## Benefits of Aquaponics

Aquaponics offers numerous benefits that make it an attractive option for homesteaders:

1. **Water Efficiency**: Aquaponics uses up to 90% less water than traditional soil-based agriculture because water is recirculated and reused within the system.

2. **Space Efficiency**: Aquaponics systems can be set up in small spaces, making them ideal for urban homesteads or areas with limited arable land.

3. **Sustainable Production**: By combining fish farming and plant cultivation, aquaponics creates a sustainable, closed-loop system

that reduces waste and the need for external inputs.

4. **Year-Round Production**: With the ability to control environmental conditions, aquaponics allows for year-round production of fresh produce and fish.

5. **High Yields**: The nutrient-rich water and controlled environment of aquaponics systems often result in faster plant growth and higher yields compared to traditional methods.

## Setting Up a Hydroponic System

Hydroponics is a method of growing plants without soil, using nutrient-rich water to deliver essential nutrients directly to the plant roots. Hydroponic systems are highly efficient and can be tailored to fit various spaces and crop types. Setting up a hydroponic system involves selecting the right type of system, preparing the necessary equipment, and managing the growing environment.

## Types of Hydroponic Systems

There are several types of hydroponic systems, each with its unique advantages and applications. The most common types include:

1. **Wick System**: The simplest form of hydroponics, a wick system uses a wick to draw nutrient solution from a reservoir into the growing medium. It is ideal for small-scale or beginner setups but may not be suitable for larger plants with high nutrient needs.

2. **Deep Water Culture (DWC)**: In DWC systems, plants are suspended above a nutrient solution with their roots submerged

in the water. Air stones provide oxygen to the roots, preventing root rot. DWC is excellent for growing leafy greens and herbs but requires careful oxygenation management.

3. **Nutrient Film Technique (NFT)**: NFT systems use a thin film of nutrient solution that flows continuously over the plant roots in a sloped channel. This system is efficient and suitable for fast-growing plants like lettuce and herbs but requires precise management of water flow and nutrient levels.

4. **Ebb and Flow (Flood and Drain)**: In an ebb and flow system, the grow bed is periodically flooded with nutrient solution and then drained back into the reservoir. This cycle provides plants with nutrients, oxygen, and moisture. Ebb and flow systems are versatile and can support a wide variety of crops.

5. **Drip System**: A drip system delivers nutrient solution directly to the plant roots through a network of drip lines and emitters. It is highly customizable and can be used for large-scale production of various crops. Drip systems require regular maintenance to prevent clogging and ensure even nutrient distribution.

6. **Aeroponics**: In aeroponics, plant roots are suspended in the air and misted with a nutrient solution. This system provides excellent oxygenation and nutrient absorption, resulting in rapid growth and high yields. However, aeroponics systems are complex and require precise control of environmental conditions.

## Setting Up Your Hydroponic System

Setting up a hydroponic system involves selecting the appropriate type, assembling the necessary equipment, and creating an optimal growing environment.

1. **Selecting the System**: Choose a hydroponic system based on your space, budget, and crop preferences. Beginners may start with simpler systems like wick or DWC, while more experienced growers can explore NFT or aeroponics.

2. **Assembling Equipment**: Gather the essential components for your hydroponic system, including:

   ○ **Grow Bed or Container**: Holds the growing medium and plants.

   ○ **Reservoir**: Stores the nutrient solution.

   ○ **Pump**: Circulates the nutrient solution (for active systems).

   ○ **Growing Medium**: Supports plant roots (e.g., rockwool, perlite, coconut coir).

   ○ **Nutrients**: Hydroponic nutrient solution tailored to your crops.

   ○ **Lighting**: Provides adequate light for plant growth (e.g., LED or fluorescent lights).

   ○ **pH and EC Meters**: Monitor the pH and electrical conductivity (EC) of the nutrient solution.

   ○ **Timers**: Automate lighting, watering, and nutrient cycles.

3. **Preparing the Growing Medium**: Depending on the chosen system, prepare the growing medium to support plant roots. For example, soak rockwool cubes in water to stabilize their pH before planting seeds or seedlings.

4. **Mixing the Nutrient Solution**: Follow the manufacturer's instructions to mix the hydroponic nutrient solution. Monitor and adjust the pH to ensure it stays within the optimal range (usually 5.5 to 6.5).

5. **Planting**: Place seeds or seedlings into the growing medium, ensuring they are securely anchored. For systems like NFT or DWC, place the plant roots into the channels or suspended above the nutrient solution.

6. **Setting Up Lighting**: Install grow lights above the plants, ensuring they provide sufficient light for photosynthesis. Adjust the height and duration of lighting based on the needs of your crops.

7. **Monitoring and Maintenance**: Regularly check the pH and EC of the nutrient solution, and adjust as necessary. Monitor plant growth, nutrient levels, and system components to ensure everything functions smoothly. Clean and sanitize the system periodically to prevent disease and algae growth.

## Integrating Aquaponics into Your Homestead

Integrating aquaponics into your homestead can enhance your food production capabilities and contribute to a more sustainable lifestyle. Aquaponics systems can be adapted to fit various scales and settings, from small indoor setups to large outdoor systems. Successful integration involves careful planning, choosing the right components, and managing the system effectively.

## Planning Your Aquaponics System

Effective planning is crucial for setting up a successful aquaponics system on your homestead. Consider the following factors during the planning process:

1. **Goals and Scale**: Define your goals for the aquaponics system. Are you looking to supplement your family's food supply, generate income, or both? Determine the scale of the system based on your goals, available space, and resources.

2. **Location**: Choose a suitable location for your aquaponics system. Indoor systems offer better control over environmental conditions, while outdoor systems can take advantage of natural sunlight. Consider factors such as temperature, light availability, and access to water and electricity.

3. **System Design**: Decide on the type of aquaponics system that best suits your needs and space. Media-based systems are ideal for beginners, while NFT and DWC systems offer higher efficiency for experienced growers. Design the layout to optimize space and ensure easy access for maintenance.

4. **Budget**: Determine your budget for setting up and maintaining the aquaponics system. Consider the costs of equipment, fish, plants, and ongoing expenses such as electricity and fish feed. Look for cost-saving opportunities, such as repurposing materials or building DIY components.

## Setting Up Your Aquaponics System

Once you have a plan in place, follow these steps to set up your aquaponics system:

1. **Assemble Components**: Gather the essential components for your aquaponics system, including:

   - **Fish Tank**: A tank or container to house the fish. Choose a size that matches the scale of your system and the number of fish you plan to raise.

   - **Grow Bed**: A container filled with a growing medium to support the plants. Ensure it is sturdy and has proper drainage.

   - **Water Pump**: A pump to circulate water between the fish tank and the grow bed. Choose a pump with the appropriate flow rate for your system.

   - **Aeration System**: An air pump and air stones to provide oxygen to the fish and plant roots.

   - **Plumbing**: Pipes, fittings, and valves to connect the components and ensure smooth water flow.

   - **Fish**: Select fish species that are well-suited to aquaponics, such as tilapia, trout, or catfish.

   - **Plants**: Choose plants that thrive in aquaponic systems, such as leafy greens, herbs, and vegetables.

2. **Set Up the Fish Tank**: Place the fish tank in a suitable location and

fill it with water. Install the aeration system to provide oxygen to the fish. Allow the water to circulate for a few days to ensure it is properly aerated and conditioned.

3. **Prepare the Grow Bed**: Fill the grow bed with a growing medium, such as gravel or clay pebbles. Ensure the medium is thoroughly rinsed to remove dust and debris. Install the water pump and plumbing to circulate water from the fish tank to the grow bed.

4. **Cycle the System**: Before adding fish and plants, cycle the system to establish beneficial bacteria that convert fish waste into nitrates. This process, known as "fishless cycling," involves adding a source of ammonia to the fish tank and monitoring water parameters until ammonia and nitrite levels drop to zero and nitrates are present. This typically takes 4-6 weeks.

5. **Add Fish and Plants**: Once the system is cycled, introduce fish to the tank gradually to avoid overloading the biofilter. Start with a few fish and monitor water parameters closely. Add plants to the grow bed, ensuring their roots are well-anchored in the growing medium.

6. **Monitor and Maintain**: Regularly monitor water quality, including pH, ammonia, nitrite, and nitrate levels. Adjust feeding rates, water flow, and system components as needed to maintain optimal conditions. Perform routine maintenance, such as cleaning filters, checking for leaks, and removing dead plant material.

## Managing and Optimizing Your Aquaponics System

Effective management and optimization are essential for maximizing the productivity and sustainability of your aquaponics system. Follow these best practices to ensure your system thrives:

1. **Water Quality Management**: Maintain optimal water quality by regularly testing and adjusting parameters. Keep pH levels between 6.8 and 7.2, and ensure ammonia and nitrite levels remain low. Use buffers to stabilize pH and add fresh water as needed to replace evaporated or used water.

2. **Fish Health and Feeding**: Monitor fish health closely, watching for signs of stress, disease, or poor growth. Feed fish a balanced diet appropriate for their species and size, and avoid overfeeding to prevent water quality issues. Remove uneaten food and waste promptly.

3. **Plant Care**: Ensure plants receive adequate light, nutrients, and space to grow. Prune plants regularly to promote healthy growth and prevent overcrowding. Monitor for pests and diseases, and use organic pest control methods to protect your plants.

4. **System Maintenance**: Perform regular maintenance to keep your system running smoothly. Clean filters, check and adjust water flow, inspect plumbing for leaks, and replace worn or damaged components. Periodically clean the grow bed to remove accumulated solids and maintain good water flow.

5. **Harvesting and Replanting**: Harvest fish and plants as they reach maturity, ensuring a continuous supply of fresh produce. Stagger plantings to maintain a consistent harvest and prevent gaps in production. After harvesting, replant immediately to keep the

system balanced and productive.

## Scaling Up and Expanding Your System

As you gain experience and confidence with your aquaponics system, you may choose to scale up and expand your operation. Consider these strategies for expanding your aquaponics system:

1. **Adding More Grow Beds**: Increase your growing capacity by adding additional grow beds to your existing system. Ensure your water pump and biofilter can handle the increased load, and distribute water evenly to all grow beds.

2. **Introducing New Crop Varieties**: Experiment with new crop varieties to diversify your production and increase yield. Consider adding fruiting plants, such as tomatoes and peppers, which may require additional support structures and nutrients.

3. **Expanding Fish Production**: Increase fish production by adding more fish tanks or raising higher-value fish species. Ensure you have adequate filtration, aeration, and space to support the additional fish.

4. **Vertical Farming**: Utilize vertical space to maximize production in limited areas. Vertical farming techniques, such as tower gardens or stacked grow beds, can significantly increase your growing capacity without expanding your footprint.

5. **Greenhouse Integration**: Integrate your aquaponics system into a greenhouse to extend the growing season and protect plants from adverse weather conditions. A greenhouse provides a controlled

environment that enhances plant growth and fish health.

Aquaponics and hydroponics offer innovative and sustainable solutions for maximizing food production on your homestead. By understanding the basics of these systems, setting up and managing your hydroponic system, and integrating aquaponics into your homestead, you can achieve higher yields, greater efficiency, and improved sustainability. Embrace these advanced techniques to take your homesteading to the next level and enjoy the benefits of fresh, homegrown produce year-round. With careful planning, diligent management, and a commitment to continuous learning, you can create a thriving and resilient farming system that supports your family's needs and contributes to a more sustainable future.

# Chapter 20

# Wild Edibles and Foraging

Foraging for wild edibles connects us to our ancestral roots, deepens our appreciation for nature, and adds a diverse range of nutritious foods to our diet. This chapter delves into the art of identifying edible wild plants, ensuring safety and sustainability in foraging, and integrating wild edibles into your diet to enhance both your health and your culinary repertoire.

## Identifying Edible Wild Plants

The first step in foraging is learning to accurately identify edible wild plants. This requires knowledge, practice, and a cautious approach to avoid poisonous look-alikes and to ensure that you harvest plants responsibly.

## Common Edible Wild Plants

1. **Dandelion (Taraxacum officinale):**

   ○ **Identification**: Dandelions have bright yellow flowers and jagged, tooth-like leaves. The plant produces a single flower per stem, which turns into a white, fluffy seed head.

- **Edible Parts**: Leaves, flowers, and roots.

- **Uses**: Leaves can be eaten raw in salads or cooked like spinach. Flowers can be used in teas, salads, or to make dandelion wine. Roots can be roasted and used as a coffee substitute.

## 2. Chickweed (Stellaria media):

- **Identification**: Chickweed has small, star-shaped white flowers and smooth, oval leaves. It grows in a sprawling mat and has a single line of fine hairs on one side of the stem.

- **Edible Parts**: Leaves, stems, and flowers.

- **Uses**: Can be eaten raw in salads or cooked as a leafy green. Chickweed is also used in herbal remedies for its anti-inflammatory properties.

## 3. Nettle (Urtica dioica):

- **Identification**: Nettles have serrated, heart-shaped leaves and are covered in tiny, stinging hairs. They produce clusters of small, greenish flowers.

- **Edible Parts**: Young leaves.

- **Uses**: Leaves must be cooked or dried to remove the stinging effect. They can be used in soups, teas, and as a substitute for spinach.

4. **Wild Garlic (Allium ursinum)**:

- **Identification**: Wild garlic has broad, lance-shaped leaves, white star-shaped flowers, and a strong garlic scent.

- **Edible Parts**: Leaves, flowers, and bulbs.

- **Uses**: Leaves can be used raw in salads, pestos, or as a seasoning. Flowers are edible and add a decorative touch to dishes. Bulbs can be used like regular garlic.

5. **Plantain (Plantago major and Plantago lanceolata)**:

- **Identification**: Plantain has broad, ribbed leaves that grow in a rosette pattern. Plantago major has wider leaves, while Plantago lanceolata has narrower, lance-shaped leaves.

- **Edible Parts**: Leaves and seeds.

- **Uses**: Leaves can be eaten raw in salads or cooked. Seeds can be ground into flour or used as a fiber supplement.

6. **Wood Sorrel (Oxalis spp.)**:

- **Identification**: Wood sorrel has clover-like leaves and small, yellow, white, or pink flowers. It grows low to the ground and has a tangy, lemony flavor.

- **Edible Parts**: Leaves, flowers, and seed pods.

- **Uses**: Leaves and flowers can be eaten raw in salads or used as a garnish. The tangy flavor adds a refreshing twist to dishes.

7. **Wild Asparagus (Asparagus officinalis):**

   ○ **Identification**: Wild asparagus looks similar to cultivated asparagus but is often thinner and more spread out. It has feathery, fern-like foliage.

   ○ **Edible Parts**: Young shoots.

   ○ **Uses**: Shoots can be harvested in spring and eaten raw, steamed, or sautéed, just like cultivated asparagus.

8. **Wild Berries (Rubus spp., Vaccinium spp., Sambucus spp.):**

   ○ **Identification**: Various species include blackberries, raspberries, blueberries, and elderberries. Identify them by their distinct fruits, leaves, and growth habits.

   ○ **Edible Parts**: Berries.

   ○ **Uses**: Berries can be eaten fresh, dried, or used in jams, jellies, pies, and beverages.

## Field Guides and Resources

Using field guides and resources is essential for accurate plant identification and foraging success. Reliable guides provide detailed descriptions, photos, and information about edible parts, uses, and potential look-alikes.

1. **Books**: Invest in comprehensive foraging books specific to your region. Some highly recommended titles include:

   ○ "The Forager's Harvest" by Samuel Thayer

- "Edible Wild Plants: Wild Foods from Dirt to Plate" by John Kallas

- "The Forager's Guide to Wild Foods" by Nicole Apelian

2. **Apps**: Several mobile apps can aid in plant identification and provide valuable information on wild edibles. Popular apps include:

   - PlantSnap

   - iNaturalist

   - Wild Edibles Forage

3. **Local Experts**: Join local foraging groups, attend workshops, and connect with experienced foragers who can provide hands-on guidance and share their knowledge.

4. **Herbariums and Botanical Gardens**: Visit local herbariums and botanical gardens to see and study live plant specimens. These institutions often offer educational programs and guided tours focused on native plants and foraging.

# Chapter 21

# Self-Sufficiency Beyond the Garden

Achieving self-sufficiency extends beyond just growing your own food. It encompasses a broader approach to managing resources, reducing dependence on external inputs, and fostering a sustainable lifestyle. In this section, we will explore how to harness renewable energy for your farm, undertake DIY projects for increased farm efficiency, and build a supportive community network.

## Renewable Energy for Your Farm

Renewable energy sources offer a sustainable and cost-effective way to power your homestead. By reducing reliance on fossil fuels, you can lower your carbon footprint, decrease energy costs, and increase your farm's resilience. Several renewable energy options are well-suited for homesteads, including solar, wind, and biomass energy.

### Solar Energy

Solar energy is one of the most accessible and versatile renewable energy sources. Photovoltaic (PV) panels convert sunlight into electricity, which can power various farm operations, from lighting and irrigation to household needs.

1. **Assessing Solar Potential**: Before investing in solar panels, assess your location's solar potential. Consider factors such as average sunlight hours, shading from trees or buildings, and roof orientation. South-facing roofs receive the most sunlight in the Northern Hemisphere, making them ideal for solar panels.

2. **Sizing Your System**: Determine your energy needs by calculating your average daily electricity consumption. This information helps you size your solar system appropriately. For a more accurate assessment, analyze your utility bills over the past year to understand seasonal variations in energy use.

3. **Types of Solar Systems**: Choose between grid-tied, off-grid, and hybrid solar systems:

   ○ **Grid-Tied**: These systems are connected to the utility grid, allowing you to draw power when needed and sell excess electricity back to the grid. They are cost-effective and eliminate the need for battery storage.

   ○ **Off-Grid**: Off-grid systems are independent of the utility grid and require battery storage to store excess energy for use during cloudy days or nighttime. They offer complete energy independence but are more expensive due to battery costs.

   ○ **Hybrid**: Hybrid systems combine grid-tied and off-grid

features, allowing you to store excess energy in batteries while remaining connected to the grid. They provide flexibility and reliability.

4. **Installation and Maintenance**: Install solar panels on rooftops or ground-mounted structures. Ensure they are securely fastened and positioned to maximize sunlight exposure. Regularly clean the panels to remove dust and debris, and inspect the system for any signs of wear or damage.

5. **Incentives and Rebates**: Research available incentives, tax credits, and rebates to offset the initial costs of solar installation. Many governments and utilities offer financial incentives to encourage the adoption of renewable energy.

## Wind Energy

Wind energy is another viable option for homesteads located in areas with consistent and strong winds. Wind turbines convert wind energy into electricity, which can supplement or replace grid power.

1. **Assessing Wind Potential**: Evaluate your site's wind potential by measuring average wind speeds. Wind speeds of at least 9-12 miles per hour are generally required for efficient turbine operation. Use an anemometer to gather data over a period of time.

2. **Types of Wind Turbines**: Choose between horizontal-axis and vertical-axis wind turbines:

   ○ **Horizontal-Axis Wind Turbines (HAWT)**: These are the most common type, with blades that rotate around a horizontal axis. They are efficient and suitable for areas with consistent wind patterns.

- ○ **Vertical-Axis Wind Turbines (VAWT)**: VAWTs have blades that rotate around a vertical axis. They perform well in turbulent wind conditions and can be installed closer to the ground.

3. **Sizing and Placement**: Determine your energy needs and select a turbine size accordingly. Consider factors such as tower height and proximity to buildings or trees, which can cause turbulence and reduce efficiency. Place turbines in open areas with minimal obstructions.

4. **Installation and Maintenance**: Install wind turbines on sturdy towers anchored securely to the ground. Regularly inspect and maintain the turbine, checking for wear and tear on blades, bearings, and electrical components. Lubricate moving parts and replace damaged components promptly.

5. **Regulations and Permits**: Check local regulations and obtain necessary permits before installing a wind turbine. Some areas have zoning restrictions or height limits that may affect your installation plans.

## Biomass Energy

Biomass energy involves using organic materials, such as wood, crop residues, and animal manure, to generate heat or electricity. It is a versatile and sustainable energy source that can help manage waste on your farm.

1. **Types of Biomass Energy Systems**:

- **Biomass Boilers**: These systems burn organic materials to produce heat for space heating, water heating, or greenhouse heating. They are efficient and can use a variety of biomass fuels.

- **Anaerobic Digesters**: Anaerobic digesters convert organic waste into biogas through microbial digestion. The biogas can be used to generate electricity or heat, and the byproduct, digestate, can be used as fertilizer.

- **Pellet Stoves**: Pellet stoves burn compressed biomass pellets to produce heat. They are highly efficient and suitable for heating small spaces or individual rooms.

2. **Fuel Sources**: Identify available biomass resources on your farm, such as wood, crop residues, or animal manure. Ensure a consistent supply to maintain efficient operation of your biomass energy system.

3. **System Design and Installation**: Design your biomass energy system based on your energy needs and available resources. Install the system according to manufacturer guidelines, ensuring proper ventilation and safety measures.

4. **Operation and Maintenance**: Regularly maintain your biomass energy system to ensure efficient operation. Clean and inspect the system, remove ash or waste products, and check for signs of wear or damage.

5. **Environmental Impact**: Use sustainable practices when sourcing biomass to minimize environmental impact. Avoid overharvesting wood or depleting crop residues, and manage animal waste

responsibly.

## DIY Projects for Farm Efficiency

Undertaking DIY projects can significantly improve the efficiency and sustainability of your farm. These projects often require minimal investment but offer substantial returns in terms of resource conservation, productivity, and self-sufficiency.

## Rainwater Harvesting

Rainwater harvesting involves collecting and storing rainwater for agricultural and household use. It is an effective way to conserve water, reduce reliance on municipal supplies, and ensure a steady water supply during dry periods.

1. **Gutter and Downspout System**: Install gutters and downspouts on rooftops to capture rainwater. Ensure they are clean and free of debris to maximize water collection.

2. **Storage Tanks**: Choose storage tanks based on your water needs and available space. Options include above-ground barrels, underground cisterns, or large tanks. Ensure tanks are made of food-grade materials and have proper filtration to keep water clean.

3. **Filtration and Purification**: Install filters and purification systems to remove debris, sediments, and contaminants from collected rainwater. First-flush diverters can be used to discard the initial runoff, which may contain higher levels of pollutants.

4. **Distribution System**: Set up a distribution system to deliver stored

rainwater to your garden, livestock, or household. Use gravity-fed systems, pumps, or drip irrigation to ensure efficient water delivery.

5. **Maintenance**: Regularly inspect and clean your rainwater harvesting system to ensure optimal performance. Check for leaks, clean filters, and remove debris from gutters and downspouts.

## Composting Systems

Composting is a sustainable way to recycle organic waste into nutrient-rich compost for your garden. It reduces waste, improves soil fertility, and enhances plant growth.

1. **Compost Bins or Piles**: Choose a composting method based on your space and needs. Options include compost bins, tumblers, or open piles. Ensure the composting site is well-drained and receives adequate airflow.

2. **Materials**: Collect a balanced mix of green (nitrogen-rich) and brown (carbon-rich) materials. Green materials include kitchen scraps, grass clippings, and manure, while brown materials include leaves, straw, and cardboard.

3. **Layering and Aeration**: Layer green and brown materials to create a balanced compost pile. Turn the pile regularly to aerate it and speed up decomposition. Maintain moisture levels by adding water if the pile becomes too dry.

4. **Temperature Monitoring**: Monitor the temperature of your compost pile to ensure it stays within the optimal range (135-160°F) for decomposition. High temperatures indicate active microbial

activity and help kill pathogens and weed seeds.

5. **Harvesting Compost**: Compost is ready to use when it is dark, crumbly, and has an earthy smell. Use finished compost to enrich garden soil, improve soil structure, and enhance plant growth.

## Solar Dryers

Solar dryers use the sun's energy to dry fruits, vegetables, herbs, and other produce. They are a cost-effective and energy-efficient way to preserve food.

1. **Design and Construction**: Build a solar dryer using readily available materials. Designs vary, but most include a transparent cover to trap heat, a dark-colored interior to absorb heat, and shelves or racks for drying produce. Ensure proper ventilation to allow moisture to escape.

2. **Location**: Place the solar dryer in a sunny, well-ventilated area. Ensure it is positioned to receive maximum sunlight throughout the day.

3. **Preparation of Produce**: Wash and slice produce uniformly to ensure even drying. Blanch vegetables if necessary to preserve color and texture. Arrange produce in a single layer on drying racks to allow airflow.

4. **Drying Process**: Monitor the drying process regularly. Turn or rotate produce to ensure even drying. Drying times vary depending on the type of produce, thickness, and weather conditions. Most fruits and vegetables take 1-3 days to dry completely.

5. **Storage**: Store dried produce in airtight containers in a cool, dark place. Properly dried and stored produce can last for several months to a year.

## Root Cellars

A root cellar is a cool, dark, and humid storage space that extends the shelf life of root vegetables, fruits, and other perishable items. It is a traditional and effective way to preserve food without refrigeration.

1. **Site Selection**: Choose a location for your root cellar that stays cool and has good drainage. Underground or partially underground locations are ideal for maintaining stable temperatures and humidity levels.

2. **Design and Construction**: Build a root cellar using materials such as wood, concrete, or stone. Ensure it has proper insulation, ventilation, and drainage. Install shelves or bins to store produce, and maintain a temperature range of 32-40°F with humidity levels around 85-95%.

3. **Preparation and Storage**: Harvest produce at peak maturity and handle it carefully to avoid bruising. Cure vegetables like potatoes, garlic, and onions before storing them to extend their shelf life. Store produce in layers, leaving space for air circulation.

4. **Monitoring and Maintenance**: Regularly check the temperature and humidity levels in your root cellar. Adjust ventilation as needed to maintain optimal conditions. Inspect stored produce for signs of spoilage and remove any affected items promptly.

## Building a Community Network

Building a strong community network is essential for achieving self-sufficiency and sustainability. A supportive community can provide resources, knowledge, and assistance, enhancing your homesteading efforts and fostering resilience.

## Joining Local Farming Groups

Local farming groups offer a platform for networking, learning, and collaboration. Joining these groups can provide valuable insights, resources, and support.

1. **Finding Groups**: Look for local farming groups, cooperatives, or organizations focused on sustainable agriculture. Check community centers, agricultural extension offices, or online platforms for information.

2. **Participation**: Attend meetings, workshops, and events organized by these groups. Actively participate in discussions, share your experiences, and seek advice from experienced farmers.

3. **Collaboration**: Collaborate with group members on projects, such as bulk purchasing of supplies, shared use of equipment, or joint marketing efforts. Pooling resources can reduce costs and increase efficiency.

## Networking with Neighbors

Building relationships with your neighbors can create a supportive and cooperative community. Neighbors can offer assistance, share resources, and provide a sense of security.

1. **Introduction and Communication**: Introduce yourself to your neighbors and establish open lines of communication. Share your goals and interests in homesteading and learn about their skills and resources.

2. **Resource Sharing**: Exchange tools, equipment, and knowledge with neighbors. For example, you can share gardening tools, trade seeds, or exchange tips on pest management.

3. **Community Projects**: Organize community projects, such as neighborhood cleanups, tree planting, or community gardens. These projects can foster a sense of community and collective responsibility.

4. **Emergency Preparedness**: Collaborate with neighbors to develop emergency preparedness plans. Share information about resources, such as water sources, medical supplies, and shelter, to ensure mutual support during emergencies.

## Participating in Farmers Markets

Farmers markets provide an excellent opportunity to connect with other local farmers, sell your produce, and engage with the community. Participating in farmers markets can enhance your visibility and build relationships with customers.

1. **Market Research**: Research local farmers markets to identify those

that align with your products and values. Visit markets to observe the setup, customer base, and vendor offerings.

2. **Preparation**: Prepare for market participation by ensuring you have high-quality produce, attractive packaging, and marketing materials. Invest in a professional-looking booth setup, including tables, tents, and signage.

3. **Engagement**: Engage with customers by sharing your farming practices, offering samples, and providing recipes or cooking tips. Building relationships with customers can lead to repeat business and word-of-mouth referrals.

4. **Networking with Vendors**: Network with other vendors to share experiences, exchange tips, and explore collaboration opportunities. Building a supportive network of fellow farmers can provide valuable insights and assistance.

## Online Communities and Resources

Online communities and resources offer a wealth of information, support, and networking opportunities for homesteaders. Joining online forums, social media groups, and educational platforms can enhance your knowledge and connect you with like-minded individuals.

1. **Finding Online Communities**: Search for online forums, social media groups, and websites dedicated to homesteading, sustainable agriculture, and renewable energy. Popular platforms include Facebook, Reddit, and specialized farming websites.

2. **Active Participation**: Actively participate in online discussions by

asking questions, sharing your experiences, and providing advice to others. Engaging with the community can build relationships and expand your knowledge.

3. **Accessing Resources**: Utilize online resources, such as articles, videos, and webinars, to learn about various aspects of homesteading. Many organizations and educational institutions offer free or affordable online courses and workshops.

4. **Networking Opportunities**: Use online platforms to network with other homesteaders, farmers, and experts. Join virtual events, attend webinars, and participate in online conferences to expand your network and gain new insights.

Self-sufficiency beyond the garden involves harnessing renewable energy, undertaking DIY projects, and building a strong community network. By incorporating renewable energy sources like solar, wind, and biomass, you can reduce your reliance on external energy inputs and enhance your farm's sustainability. DIY projects, such as rainwater harvesting, composting, solar drying, and building root cellars, can improve efficiency and resource conservation on your homestead. Building a supportive community network through local farming groups, neighbor collaborations, farmers markets, and online communities can provide valuable resources, knowledge, and support, enhancing your self-sufficiency and resilience. Embrace these strategies to create a sustainable, efficient, and interconnected homestead that thrives beyond the garden.

# Chapter 22

# Permaculture Principles

Permaculture is a holistic approach to designing sustainable and self-sufficient agricultural systems. It integrates land, resources, people, and the environment through mutually beneficial synergies. By mimicking natural ecosystems, permaculture principles can be applied to homesteads to enhance productivity, sustainability, and resilience. This section will introduce permaculture, guide you through designing a permaculture garden, and outline how to implement sustainable practices effectively.

## Introduction to Permaculture

Permaculture, a term coined by Bill Mollison and David Holmgren in the 1970s, is a contraction of "permanent agriculture" and "permanent culture." It emphasizes creating agricultural systems that work with nature rather than against it, aiming to build sustainable and regenerative ecosystems.

## Core Principles of Permaculture

Permaculture is based on a set of core principles that guide design and implementation. These principles can be adapted to various environments and scales, from small gardens to large farms.

1. **Observe and Interact**: Spend time observing natural systems and interactions in your environment. Understanding how elements in nature interact helps you design systems that mimic these relationships, leading to more sustainable and efficient practices.

2. **Catch and Store Energy**: Make use of renewable resources and capture energy when it is abundant. This could include collecting rainwater, harnessing solar power, and using wind energy. Stored energy can be used in times of scarcity.

3. **Obtain a Yield**: Ensure that your system provides tangible outputs that meet your needs, such as food, fuel, fiber, and other resources. Every element in your design should contribute to the overall productivity of the system.

4. **Apply Self-Regulation and Accept Feedback**: Implement systems that can regulate themselves and adapt based on feedback. This principle encourages you to monitor your systems and make adjustments as needed to improve efficiency and sustainability.

5. **Use and Value Renewable Resources and Services**: Prioritize renewable resources over non-renewable ones. This includes using natural materials, growing renewable crops, and incorporating animals in ways that enhance the ecosystem.

6. **Produce No Waste**: Design systems that recycle and reuse waste. Every output should become an input for another process, creating

a closed-loop system that minimizes waste.

7. **Design from Patterns to Details**: Observe natural patterns and use them as a foundation for your designs. Once the broad patterns are understood, you can focus on the details to fine-tune your system.

8. **Integrate Rather Than Segregate**: Place elements so that they work together to support and benefit each other. Integrated systems are more efficient and resilient.

9. **Use Small and Slow Solutions**: Start with small, manageable projects and scale up gradually. Small solutions are easier to maintain and adjust, reducing the risk of failure.

10. **Use and Value Diversity**: Biodiversity enhances resilience and productivity. A diverse system is more adaptable to changes and less vulnerable to pests and diseases.

11. **Use Edges and Value the Marginal**: Edges, or the interfaces between different elements, are often the most productive and diverse areas. Utilize these spaces effectively to maximize productivity.

12. **Creatively Use and Respond to Change**: Be adaptable and innovative in response to changes. Challenges can present opportunities for improvement and growth.

## Designing a Permaculture Garden

Designing a permaculture garden involves applying these principles to create a sustainable and productive ecosystem. The design process includes site assessment, layout planning, and selecting appropriate plants and elements.

## Site Assessment

Before designing your permaculture garden, conduct a thorough site assessment to understand the natural characteristics and potential of your land. This includes evaluating soil quality, climate, water sources, and existing vegetation.

1. **Soil Quality**: Test your soil to determine its texture, structure, pH, and nutrient content. Healthy soil is crucial for plant growth and overall ecosystem health. Amend the soil as needed to improve fertility and structure.

2. **Climate and Microclimates**: Understand the climate of your region, including temperature ranges, precipitation patterns, and frost dates. Identify microclimates within your site, such as sunny spots, shaded areas, and windbreaks, which can influence plant growth.

3. **Water Sources and Management**: Identify natural water sources, such as streams, ponds, and rainwater runoff. Plan for water management strategies, including rainwater harvesting, irrigation systems, and swales to capture and distribute water efficiently.

4. **Topography and Land Features**: Analyze the topography of your land, including slopes, elevation changes, and natural contours. Use this information to design terraces, raised beds, and other features that work with the landscape.

5. **Existing Vegetation**: Take note of existing trees, shrubs, and plants. Preserve beneficial vegetation that can provide shade, wind protection, or habitat for wildlife. Remove invasive species that compete with desired plants.

## Layout Planning

Once you have assessed your site, create a detailed layout plan that integrates the permaculture principles. The layout should maximize efficiency, productivity, and sustainability.

1. **Zones**: Divide your garden into zones based on the frequency of use and maintenance needs. Zones help organize your space efficiently and ensure that elements are placed for optimal interaction.

   ○ **Zone 0**: The home or center of activity.

   ○ **Zone 1**: Areas closest to the home, requiring frequent attention, such as kitchen gardens and herb beds.

   ○ **Zone 2**: Areas requiring regular but less frequent maintenance, such as vegetable gardens and small livestock.

   ○ **Zone 3**: Areas for larger-scale crops and orchards, needing occasional maintenance.

   ○ **Zone 4**: Semi-wild areas for foraging, timber, and wildlife habitat.

   ○ **Zone 5**: Natural wilderness left undisturbed to support biodiversity.

2. **Sectors**: Consider external factors, such as wind, sun, water flow, and views, which can influence your design. Use sectors to analyze these elements and incorporate them into your layout. For example, place windbreaks in areas prone to strong winds and position sun-loving plants in the sunniest spots.

3. **Elements and Connections**: Identify the key elements of your garden, such as plants, animals, water features, and structures. Plan their placement to create beneficial interactions. For instance, plant nitrogen-fixing plants near nutrient-demanding crops and place compost bins close to the garden for easy access.

4. **Polycultures and Guilds**: Design plant communities, or guilds, that work together to support each other. Polycultures mimic natural ecosystems and enhance biodiversity. A typical example is the "Three Sisters" planting of corn, beans, and squash, where each plant benefits the others.

5. **Paths and Access**: Design paths to provide easy access to all areas of your garden. Ensure paths are wide enough for wheelbarrows and other tools. Use materials such as gravel, wood chips, or stepping stones to create durable and permeable paths.

## Selecting Plants and Elements

Choose plants and elements that suit your site conditions, meet your needs, and contribute to the overall sustainability of your permaculture garden.

1. **Edible Plants**: Prioritize plants that provide food for your household, including fruits, vegetables, herbs, and nuts. Choose varieties that are well-suited to your climate and soil conditions.

2. **Perennials and Annuals**: Incorporate a mix of perennial and annual plants. Perennials provide long-term stability and reduce the need for replanting, while annuals offer seasonal variety and flexibility.

3. **Beneficial Plants**: Include plants that attract pollinators, repel pests, fix nitrogen, and provide mulch or compost material. Examples include comfrey, yarrow, marigold, and clover.

4. **Animals**: Integrate animals into your permaculture system to provide manure, pest control, and other benefits. Chickens, ducks, bees, and rabbits are common choices for small-scale permaculture gardens.

5. **Water Features**: Add ponds, swales, and rain gardens to manage water effectively and create habitats for beneficial wildlife. Water features also enhance the aesthetic appeal of your garden.

6. **Structures**: Incorporate structures such as greenhouses, sheds, trellises, and cold frames to extend the growing season, protect plants, and support climbing crops. Use sustainable materials and designs that blend with the natural environment.

## Implementing Sustainable Practices

Implementing sustainable practices in your permaculture garden ensures long-term productivity and ecological balance. These practices include soil management, water conservation, integrated pest management, and energy-efficient systems.

## Soil Management

Healthy soil is the foundation of a productive permaculture garden. Implement practices that build and maintain soil fertility and structure.

1. **Composting**: Create compost from kitchen scraps, garden waste, and animal manure. Compost enriches the soil with organic matter, improves soil structure, and provides essential nutrients. Use compost bins, piles, or tumblers to produce high-quality compost.

2. **Mulching**: Apply mulch to garden beds to retain moisture, suppress weeds, and add organic matter to the soil. Use materials such as straw, wood chips, leaves, or grass clippings. Mulching also moderates soil temperature and reduces erosion.

3. **Cover Cropping**: Plant cover crops during off-seasons to protect and enrich the soil. Cover crops, such as clover, vetch, and rye, prevent erosion, fix nitrogen, and add organic matter. Incorporate cover crops into the soil before planting main crops.

4. **No-Till Gardening**: Avoid tilling the soil to preserve its structure and microbial life. No-till gardening reduces soil erosion, improves water retention, and promotes healthier plants. Use mulch and cover crops to suppress weeds and maintain soil fertility.

5. **Crop Rotation**: Rotate crops to prevent nutrient depletion and reduce pest and disease pressure. Different crops have varying nutrient needs and pest associations. Rotating crops disrupts pest cycles and maintains soil health.

## Water Conservation

Efficient water management is crucial for a sustainable permaculture garden. Implement practices that conserve water and ensure its effective use.

1. **Rainwater Harvesting**: Collect and store rainwater for irrigation. Install gutters, downspouts, and storage tanks to capture rainwater from rooftops. Use the stored water for garden beds, livestock, and household needs.

2. **Drip Irrigation**: Use drip irrigation systems to deliver water directly to plant roots. Drip irrigation reduces water waste, minimizes evaporation, and ensures plants receive consistent moisture. Install timers to automate watering schedules.

3. **Swales and Contour Planting**: Create swales and plant along contour lines to capture and slow down water runoff. Swales are shallow trenches that follow the natural contours of the land, allowing water to infiltrate the soil and recharge groundwater.

4. **Greywater Recycling**: Reuse greywater from sinks, showers, and washing machines for irrigation. Ensure greywater is free of harmful chemicals and detergents. Use simple filtration systems to treat greywater before applying it to garden beds.

5. **Drought-Resistant Plants**: Choose drought-resistant and native plants that require less water. These plants are adapted to local conditions and are more resilient during dry periods.

## Integrated Pest Management (IPM)

Integrated Pest Management (IPM) combines various strategies to manage pests in an environmentally friendly and sustainable way. IPM focuses on prevention, monitoring, and control.

1. **Prevention**: Implement cultural practices that reduce pest habitats and encourage healthy plant growth. Rotate crops, use clean seeds and transplants, and maintain garden cleanliness to prevent pest infestations.

2. **Beneficial Insects**: Attract and protect beneficial insects that prey on pests. Plant flowers and herbs that provide nectar and habitat for predators and parasitoids, such as ladybugs, lacewings, and parasitic wasps.

3. **Biological Control**: Use biological control agents, such as predatory insects, nematodes, and microbial pesticides, to manage pest populations. These agents are natural enemies of pests and help maintain ecological balance.

4. **Mechanical Control**: Employ physical methods to remove or exclude pests. Handpick pests, use barriers such as row covers, and install traps to catch and monitor pest activity.

5. **Organic Pesticides**: Use organic and natural pesticides as a last resort. These products are less harmful to beneficial insects and the environment. Examples include neem oil, insecticidal soap, and diatomaceous earth.

## Energy-Efficient Systems

Incorporating energy-efficient systems into your permaculture garden reduces reliance on external energy sources and enhances sustainability.

1. **Passive Solar Design**: Use passive solar design principles to capture and utilize solar energy. Orient greenhouses, cold frames, and other structures to maximize sunlight exposure. Use thermal mass, such as stone or water, to store and release heat.

2. **Renewable Energy**: Integrate renewable energy sources, such as solar panels, wind turbines, and biomass energy, to power your homestead. Renewable energy reduces carbon emissions and lowers energy costs.

3. **Efficient Lighting**: Use energy-efficient lighting, such as LED or fluorescent bulbs, for indoor growing spaces. These lights provide adequate light for plant growth while consuming less energy.

4. **Insulation and Ventilation**: Insulate greenhouses and other structures to maintain stable temperatures and reduce energy use. Ensure proper ventilation to regulate temperature and humidity and prevent overheating.

5. **Energy-Efficient Appliances**: Choose energy-efficient appliances and tools for your homestead. Look for Energy Star-rated products and use manual or solar-powered tools whenever possible.

Permaculture principles offer a comprehensive approach to creating sustainable and productive agricultural systems. By understanding and applying these principles, designing a permaculture garden, and implementing sustainable practices, you can enhance the resilience and efficiency of your homestead. Embrace permaculture to build a self-sufficient, environmentally friendly, and thriving homestead that works in harmony with nature. With careful planning, continuous learning, and a commitment to sustainability, you can achieve a permaculture system that meets your needs and contributes to the health of the planet.

# Chapter 23

# Mushroom Cultivation

Mushroom cultivation is a rewarding and productive addition to any homestead. Not only do mushrooms provide a nutritious and delicious food source, but they also play a crucial role in ecological balance and soil health. This section will delve into the benefits of growing mushrooms, the steps to set up a mushroom growing area, and the techniques for harvesting and using mushrooms.

## Benefits of Growing Mushrooms

Growing mushrooms offers numerous advantages that enhance both your homestead's sustainability and your personal well-being. These benefits range from nutritional value to environmental impacts and economic opportunities.

## Nutritional and Culinary Benefits

Mushrooms are a highly nutritious food, rich in essential vitamins, minerals, and antioxidants. They are low in calories, fat, and cholesterol, making them an excellent addition to a healthy diet.

1. **Vitamins and Minerals**: Mushrooms are a good source of B vitamins, including riboflavin, niacin, and pantothenic acid, which are vital for energy metabolism. They also provide important minerals like selenium, copper, potassium, and phosphorus.

2. **Antioxidants**: Mushrooms contain powerful antioxidants, such as ergothioneine and glutathione, which help protect cells from damage and boost the immune system.

3. **Dietary Fiber**: Mushrooms contribute dietary fiber, which supports digestive health and helps maintain a healthy weight.

4. **Culinary Versatility**: Mushrooms are versatile in the kitchen, enhancing a wide range of dishes with their unique flavors and textures. They can be sautéed, grilled, roasted, or added to soups, stews, and salads.

## Environmental Benefits

Mushroom cultivation can positively impact the environment, promoting sustainability and ecological balance.

1. **Waste Recycling**: Mushrooms can be grown on various organic substrates, such as agricultural waste, straw, and wood chips. This recycling process reduces waste and converts it into valuable food and compost.

2. **Soil Health**: Mushrooms play a critical role in decomposing organic matter, enriching the soil with nutrients, and improving its structure. They contribute to a healthy soil ecosystem by fostering beneficial microbial activity.

3. **Low Water and Energy Use**: Compared to traditional agriculture, mushroom cultivation requires relatively low water and energy inputs, making it an efficient and sustainable farming practice.

## Economic Benefits

Growing mushrooms can provide economic benefits by offering additional income streams for your homestead.

1. **High-Value Crop**: Specialty mushrooms, such as shiitake, oyster, and lion's mane, can fetch high prices in local markets, gourmet restaurants, and health food stores.

2. **Year-Round Production**: Mushrooms can be cultivated year-round in controlled environments, providing a continuous source of income and food.

3. **Low Space Requirement**: Mushrooms can be grown in small spaces, making them suitable for urban homesteads or areas with limited land. Vertical farming techniques can further maximize production in confined spaces.

## Setting Up a Mushroom Growing Area

Creating a suitable environment for mushroom cultivation involves selecting appropriate species, preparing growing substrates, and managing environmental conditions.

## Choosing Mushroom Species

The first step in setting up a mushroom growing area is selecting the species you want to cultivate. Consider factors such as climate, available space, and market demand when choosing mushroom species.

1. **Oyster Mushrooms (Pleurotus spp.)**: Oyster mushrooms are among the easiest to grow and are well-suited for beginners. They thrive on a variety of substrates, including straw, coffee grounds, and cardboard. They grow quickly and are highly productive.

2. **Shiitake Mushrooms (Lentinula edodes)**: Shiitake mushrooms are popular for their rich flavor and medicinal properties. They are typically grown on hardwood logs or sawdust blocks. Shiitake cultivation requires more time and specific conditions but is highly rewarding.

3. **Button Mushrooms (Agaricus bisporus)**: Button mushrooms, including cremini and portobello varieties, are widely cultivated and commercially available. They are grown on composted manure and straw mixtures. Button mushrooms require precise temperature and humidity control.

4. **Lion's Mane Mushrooms (Hericium erinaceus)**: Lion's mane mushrooms are known for their unique appearance and potential health benefits. They are typically grown on hardwood sawdust or enriched substrates. Lion's mane cultivation is relatively straightforward and can be done indoors.

5. **Reishi Mushrooms (Ganoderma lucidum)**: Reishi mushrooms are valued for their medicinal properties. They are typically grown on hardwood logs or sawdust. Reishi cultivation requires patience and specific environmental conditions.

# Preparing Growing Substrates

The substrate is the material on which mushrooms grow. Preparing the right substrate is crucial for successful mushroom cultivation.

1. **Selecting Substrates**: Different mushroom species require different substrates. Common substrates include straw, wood chips, sawdust, coffee grounds, and agricultural waste. Ensure the substrate is clean, free of contaminants, and suitable for the chosen mushroom species.

2. **Pasteurization**: Pasteurize the substrate to eliminate competing organisms and create a favorable environment for mushroom mycelium. Pasteurization can be done by heating the substrate to a specific temperature (typically 160-180°F) for a set period (usually 1-2 hours).

3. **Supplementation**: Supplement the substrate with additional nutrients to enhance mushroom growth. Supplements may include bran, gypsum, or soybean meal. The type and amount of supplementation depend on the mushroom species and substrate.

4. **Inoculation**: Inoculate the pasteurized substrate with mushroom spawn (mycelium grown on a carrier material). Ensure the substrate is at the right temperature for inoculation (typically around room temperature). Mix the spawn thoroughly with the substrate to ensure even colonization.

# Environmental Conditions

Mushrooms require specific environmental conditions for optimal growth. Managing temperature, humidity, light, and air exchange is crucial for successful cultivation.

1. **Temperature**: Different mushroom species have specific temperature requirements. Most mushrooms grow best within a temperature range of 60-75°F. Monitor and control the temperature to ensure it stays within the optimal range for the chosen species.

2. **Humidity**: High humidity is essential for mushroom growth, typically around 85-95%. Use a humidifier, misting system, or manual spraying to maintain the required humidity levels. Monitor humidity with a hygrometer and adjust as needed.

3. **Light**: While some mushrooms require light for fruiting, others prefer low light conditions. For example, oyster mushrooms benefit from indirect light, while button mushrooms grow well in darkness. Provide appropriate lighting based on the species' needs.

4. **Air Exchange**: Fresh air exchange is necessary to prevent the buildup of carbon dioxide and promote healthy mushroom growth. Ensure adequate ventilation in the growing area by using fans, vents, or manually fanning the area. Monitor air quality and adjust as needed.

## Setting Up Growing Structures

Setting up the physical structures for mushroom cultivation involves choosing suitable containers, shelving, and environmental controls.

1. **Containers**: Use containers such as bags, trays, buckets, or logs, depending on the mushroom species and substrate. Ensure containers are clean, sterilized, and appropriate for the chosen cultivation method.

2. **Shelving**: Install shelving or racks to maximize space and organize the growing area. Use sturdy, easy-to-clean materials. Vertical shelving allows for efficient use of space and easier management of environmental conditions.

3. **Environmental Controls**: Set up equipment to control temperature, humidity, light, and air exchange. This may include heaters, coolers, humidifiers, lights, and fans. Use timers and sensors to automate environmental controls and maintain consistent conditions.

4. **Cleanliness and Sanitation**: Maintain a clean and sanitary growing area to prevent contamination. Regularly clean and disinfect surfaces, equipment, and containers. Use protective gear, such as gloves and masks, when handling substrates and inoculating.

## Harvesting and Using Mushrooms

Proper harvesting techniques and post-harvest handling are essential for maintaining the quality and shelf life of your mushrooms. Additionally, knowing how to use and preserve mushrooms can enhance your culinary repertoire and provide year-round benefits.

## Harvesting Mushrooms

Harvesting mushrooms at the right time and using the correct techniques ensures optimal flavor, texture, and shelf life.

1. **Timing**: Harvest mushrooms when they reach the desired size and before they begin to over-mature. Signs of readiness vary by species:

   - **Oyster Mushrooms**: Harvest when the caps are fully open but before the edges begin to curl up.

   - **Shiitake Mushrooms**: Harvest when the caps are 70-80% open and the edges are still slightly rolled under.

   - **Button Mushrooms**: Harvest when the caps are still closed or just beginning to open, depending on the desired size.

   - **Lion's Mane Mushrooms**: Harvest when the spines are elongated but still firm and white.

   - **Reishi Mushrooms**: Harvest when the fruiting body is fully formed, and the color is vibrant.

2. **Techniques**: Use clean, sharp tools such as knives or scissors to harvest mushrooms. Cut the stem at the base, close to the substrate, to avoid damaging the mycelium. Handle mushrooms gently to prevent bruising and contamination.

3. **Post-Harvest Handling**: Immediately after harvesting, clean mushrooms by brushing off dirt and debris. Avoid washing mushrooms, as excess moisture can promote spoilage. Store harvested mushrooms in breathable containers, such as paper bags or baskets, to maintain freshness.

## Using and Preserving Mushrooms

Freshly harvested mushrooms can be used in various culinary applications or preserved for later use. Proper preservation techniques ensure that mushrooms retain their quality and nutritional value.

1. **Culinary Uses**: Mushrooms are versatile and can be used in numerous recipes. Some popular cooking methods include:

   ○ **Sautéing**: Sauté mushrooms in butter or oil with garlic and herbs for a quick and flavorful side dish.

   ○ **Grilling**: Marinate mushrooms and grill them for a smoky, charred flavor. Large mushrooms, such as portobellos, can be used as a meat substitute in burgers.

   ○ **Roasting**: Toss mushrooms with olive oil, salt, and pepper, and roast them in the oven until golden and crispy.

   ○ **Soups and Stews**: Add mushrooms to soups, stews, and casseroles for added texture and umami flavor.

   ○ **Stuffing**: Stuff large mushroom caps with fillings such as cheese, breadcrumbs, and herbs for a savory appetizer.

2. **Drying**: Drying mushrooms is an effective way to preserve them for long-term storage. Use a food dehydrator, oven, or air-drying method to remove moisture from mushrooms. Once dried, store mushrooms in airtight containers in a cool, dark place. Rehydrate dried mushrooms by soaking them in water before use.

3. **Freezing**: Freezing mushrooms is another preservation method.

Clean and slice mushrooms, then blanch them in boiling water for a few minutes. Drain and cool mushrooms before spreading them on a baking sheet and freezing. Once frozen, transfer mushrooms to airtight containers or freezer bags. Use frozen mushrooms directly in cooked dishes without thawing.

4. **Canning**: Pressure canning is a safe method for preserving mushrooms. Clean and slice mushrooms, then pack them into sterilized jars. Add boiling water, leaving appropriate headspace, and process in a pressure canner according to recommended times and pressures. Store canned mushrooms in a cool, dark place.

5. **Pickling**: Pickling mushrooms adds flavor and extends their shelf life. Clean and blanch mushrooms, then pack them into sterilized jars with a pickling brine made of vinegar, water, salt, and spices. Process jars in a water bath canner for long-term storage, or refrigerate for short-term use.

6. **Fermenting**: Fermenting mushrooms is an alternative preservation method that enhances their flavor and nutritional value. Clean and slice mushrooms, then pack them into a fermentation vessel with salt and water. Allow mushrooms to ferment at room temperature for several days to weeks, depending on the desired flavor. Store fermented mushrooms in the refrigerator.

## Integrating Mushrooms into Your Homestead

Incorporating mushroom cultivation into your homestead can enhance its sustainability, productivity, and resilience. Mushrooms can be grown in various settings, including gardens, greenhouses, and indoor spaces.

## Outdoor Cultivation

Growing mushrooms outdoors allows you to utilize natural conditions and resources, such as shade, humidity, and organic materials.

1. **Garden Beds**: Incorporate mushroom cultivation into garden beds by inoculating straw, wood chips, or compost with mushroom spawn. This method works well for oyster and wine cap mushrooms. Water the beds regularly to maintain moisture levels and promote growth.

2. **Logs and Stumps**: Grow shiitake, lion's mane, and reishi mushrooms on hardwood logs or stumps. Drill holes into the wood, insert spawn, and seal with wax to protect the inoculation. Place logs in a shady, humid area and water them as needed.

3. **Mulch Layers**: Integrate mushroom spawn into mulch layers around trees, shrubs, and garden beds. This method enriches the soil, enhances plant growth, and provides a harvest of mushrooms. Oyster and wine cap mushrooms are well-suited for this approach.

4. **Natural Settings**: Utilize natural forested areas or wooded sections of your property to grow mushrooms. Scatter mushroom spawn on decaying wood, leaf litter, or composted materials. Ensure adequate shade and moisture for optimal growth.

## Indoor Cultivation

Indoor mushroom cultivation allows for year-round production and greater control over environmental conditions. It is ideal for small spaces or urban homesteads.

1. **Growing Rooms**: Set up dedicated growing rooms or areas within your home, basement, or greenhouse. Use shelving or racks to organize containers and maximize space. Control temperature, humidity, and light to create optimal growing conditions.

2. **Grow Kits**: Purchase pre-made mushroom grow kits, which include inoculated substrate and instructions. Grow kits are convenient and easy to use, making them ideal for beginners. Place kits in a suitable indoor location and follow the care guidelines.

3. **Buckets and Bags**: Grow mushrooms in buckets, bags, or other containers filled with inoculated substrate. This method is versatile and allows for easy management and harvesting. Ensure proper ventilation and humidity levels in the growing area.

4. **Hydroponic Systems**: Integrate mushrooms into hydroponic systems by using nutrient-rich water and controlled environments. This method is experimental but can offer high yields and efficient use of space.

## Conclusion

Mushroom cultivation is a valuable addition to any homestead, offering numerous benefits ranging from nutritional value to environmental sustainability and economic opportunities. By understanding the benefits of growing mushrooms, setting up a suitable growing area, and mastering harvesting and preservation techniques, you can enjoy a continuous supply of fresh, flavorful mushrooms year-round. Whether grown outdoors in garden beds and logs or indoors in dedicated growing rooms, mushrooms enhance the productivity and resilience of your homestead. Embrace the practice of mushroom cultivation to diversify your food sources, enrich your soil, and contribute to a sustainable, self-sufficient lifestyle.

# Chapter 24

# Safety and Sustainability in Foraging

Foraging can be a safe and sustainable practice if done responsibly. Ensuring safety involves proper plant identification, understanding potential hazards, and respecting the environment.

## Ensuring Safety

1. **Accurate Identification**: Never consume a plant unless you are 100% certain of its identification. Many edible plants have toxic look-alikes that can cause severe illness or death. Use multiple sources to verify plant identity and familiarize yourself with common look-alikes.

2. **Start Small**: Begin with a few easily identifiable and commonly known edible plants. Gradually expand your knowledge and repertoire as you gain confidence and experience.

3. **Avoid Contaminated Areas**: Do not forage near roads, industrial sites, or areas treated with pesticides and herbicides. Plants in these

areas may be contaminated with pollutants and harmful chemicals.

4. **Harvest Responsibly**: Take only what you need and leave enough for wildlife and plant regeneration. Overharvesting can deplete local populations and disrupt ecosystems.

5. **Know the Regulations**: Familiarize yourself with local foraging regulations and obtain necessary permits if required. Some areas have restrictions to protect endangered species and sensitive habitats.

## Potential Hazards

1. **Toxic Plants**: Be aware of toxic plants in your area. Some common toxic plants include poison hemlock, deadly nightshade, and wild parsnip. Learn to identify these plants and avoid them.

2. **Allergies and Sensitivities**: Some individuals may have allergic reactions or sensitivities to certain plants. Try small amounts of new plants to test for adverse reactions before consuming larger quantities.

3. **Food Safety**: Wash foraged plants thoroughly to remove dirt, insects, and contaminants. Some plants may require cooking or processing to neutralize toxins or improve digestibility.

## Sustainability Practices

1. **Respecting Nature**: Foraging should enhance your connection to nature and contribute to the health of ecosystems. Practice "leave no trace" principles and minimize your impact on the environment.

2. **Wildlife Consideration**: Many wild plants are vital food sources for wildlife. Be mindful of the needs of local fauna and avoid overharvesting plants that are crucial for their survival.

3. **Promoting Regrowth**: Harvest in a way that encourages plant regrowth and reproduction. For example, when harvesting roots, leave part of the root system intact to allow the plant to regenerate.

4. **Invasive Species**: Focus on harvesting invasive species where possible. Removing these plants can help restore native ecosystems and improve biodiversity.

## Integrating Wild Edibles into Your Diet

Incorporating wild edibles into your diet can enhance your meals with new flavors, textures, and nutritional benefits. From salads and soups to teas and tinctures, the possibilities are vast.

## Fresh Uses

1. **Salads and Greens**: Add fresh wild greens like dandelion leaves, chickweed, and wood sorrel to salads for a burst of flavor and nutrition. Combine with other garden greens and a light vinaigrette for a refreshing dish.

2. **Soups and Stews**: Use wild edibles such as nettle, wild garlic, and wild asparagus in soups and stews. Their unique flavors can elevate traditional recipes and add a wild twist.

3. **Pesto and Sauces**: Create unique pestos and sauces with wild herbs and greens. For example, blend wild garlic or nettle with nuts, olive

oil, and cheese to make a flavorful pesto.

4. **Herbal Teas**: Brew herbal teas with wild plants like dandelion flowers, nettle leaves, and wood sorrel. These teas can be enjoyed hot or cold and offer various health benefits.

## Cooking and Preserving

1. **Cooking Methods**: Experiment with different cooking methods to bring out the best in wild edibles. Sauté, blanch, steam, or roast plants to enhance their flavors and textures.

2. **Fermenting**: Ferment wild plants like dandelion leaves and wild garlic to create tangy and probiotic-rich foods. Fermented foods can improve gut health and add variety to your diet.

3. **Drying**: Dry wild herbs and greens for long-term storage. Use dried plants in teas, soups, and seasonings. Store dried herbs in airtight containers in a cool, dark place.

4. **Freezing**: Preserve wild greens by blanching and freezing them. Frozen greens can be used in soups, stews, and smoothies throughout the year.

5. **Infusions and Tinctures**: Make herbal infusions and tinctures with wild plants. These concentrated extracts can be used for their medicinal properties or as flavoring agents.

## Recipes

1. **Wild Greens Salad**:

- **Ingredients**: Dandelion leaves, chickweed, wood sorrel, mixed garden greens, cherry tomatoes, cucumber, red onion, olive oil, lemon juice, salt, and pepper.

- **Instructions**: Wash and chop the wild greens and garden greens. Combine in a large bowl with cherry tomatoes, cucumber, and red onion. Dress with olive oil, lemon juice, salt, and pepper. Toss well and serve.

2. **Nettle Soup**:

- **Ingredients**: Fresh nettle leaves, onion, garlic, potatoes, vegetable broth, olive oil, salt, and pepper.

- **Instructions**: Sauté chopped onion and garlic in olive oil until soft. Add diced potatoes and vegetable broth. Simmer until potatoes are tender. Add nettle leaves and cook for a few minutes until wilted. Blend the soup until smooth and season with salt and pepper.

3. **Wild Garlic Pesto**:

- **Ingredients**: Wild garlic leaves, nuts (pine nuts, walnuts, or almonds), Parmesan cheese, olive oil, lemon juice, salt, and pepper.

- **Instructions**: Blend wild garlic leaves, nuts, and Parmesan cheese in a food processor. Gradually add olive oil and lemon juice until a smooth paste forms. Season with salt and pepper. Use as a spread, dip, or pasta sauce.

4. **Dandelion Flower Fritters**:

- **Ingredients**: Dandelion flowers, flour, eggs, milk, salt, pepper, and cooking oil.

- **Instructions**: Prepare a batter by mixing flour, eggs, milk, salt, and pepper. Dip cleaned dandelion flowers in the batter and fry in hot oil until golden brown. Drain on paper towels and serve warm.

5. **Elderberry Syrup**:

- **Ingredients**: Fresh or dried elderberries, water, honey, cinnamon stick, cloves, and ginger.

- **Instructions**: Simmer elderberries, water, cinnamon, cloves, and ginger in a pot for about 30 minutes. Strain the mixture to remove solids. Allow the liquid to cool slightly and then stir in honey. Bottle the syrup and store in the refrigerator. Use as a medicinal syrup or sweetener.

## Nutritional Benefits

Incorporating wild edibles into your diet can provide a wealth of nutritional benefits. These plants are often rich in vitamins, minerals, antioxidants, and other phytonutrients.

1. **Vitamins and Minerals**: Wild edibles like dandelion greens and nettle are high in vitamins A, C, K, and B vitamins. They also provide essential minerals such as calcium, magnesium, iron, and potassium.

2. **Antioxidants**: Many wild plants are potent sources of antioxidants.

For example, elderberries and wild berries are rich in anthocyanins and flavonoids, which help protect cells from oxidative damage.

3. **Dietary Fiber**: Wild edibles contribute dietary fiber, which supports digestive health, helps regulate blood sugar levels, and promotes a feeling of fullness.

4. **Unique Phytonutrients**: Wild plants contain unique phytonutrients that may offer additional health benefits. For example, wild garlic contains allicin, which has antimicrobial and cardiovascular benefits.

Foraging for wild edibles is a practice that reconnects us with nature, enriches our diet, and supports sustainable living. By learning to identify edible wild plants, practicing safety and sustainability in foraging, and integrating wild edibles into your meals, you can enjoy a diverse and nutritious array of foods while fostering a deeper connection to the environment. Whether you are harvesting dandelions for a fresh salad, making nettle soup, or creating elderberry syrup, foraging adds a unique and rewarding dimension to your homesteading journey. Embrace the wild bounty that nature offers and enhance your health, culinary creativity, and sustainability through the art of foraging.

# Chapter 25

# **Conclusion**

Embarking on a journey to self-sufficiency is an enriching, challenging, and profoundly rewarding endeavor. As you strive to achieve greater independence and sustainability on your homestead, you will undoubtedly encounter various milestones and obstacles. This conclusion aims to provide a reflective space for your progress, offer guidance for future planning, and encourage ongoing learning and growth.

## Your Journey to Self-Sufficiency

Your journey to self-sufficiency begins with a commitment to living in harmony with nature and harnessing its resources responsibly. It involves embracing a lifestyle that prioritizes sustainability, resilience, and community. Whether you are just starting or have been on this path for years, each step forward brings you closer to a more self-reliant and fulfilling life.

## The Initial Steps

1. **Vision and Goals**: Reflect on the vision and goals that inspired you to pursue self-sufficiency. Perhaps you sought to grow your food,

reduce your carbon footprint, or build a stronger connection with the land. These initial motivations provide a foundation for your journey.

2. **Skill Acquisition**: As you began, you likely immersed yourself in learning essential skills such as gardening, animal husbandry, preserving food, and managing resources. These skills are the bedrock of self-sufficiency, enabling you to produce and sustain your own needs.

3. **Resource Management**: Efficiently managing resources such as water, soil, and energy is crucial. You may have implemented rainwater harvesting, composting, and renewable energy systems to optimize resource use and reduce waste.

4. **Community Building**: Building connections with like-minded individuals and local communities has probably been an integral part of your journey. Sharing knowledge, resources, and support strengthens your homesteading efforts and fosters a sense of belonging.

## Achievements and Milestones

Reflect on the milestones and achievements you have reached so far. These accomplishments, whether big or small, represent significant progress on your path to self-sufficiency.

1. **Successful Harvests**: Celebrate the first time you harvested your own vegetables, fruits, or herbs. Each successful growing season is a testament to your hard work and dedication.

2. **Sustainable Practices**: Implementing sustainable practices such as crop rotation, permaculture principles, and renewable energy solutions demonstrates your commitment to environmental stewardship.

3. **Skill Mastery**: Achieving proficiency in skills such as beekeeping, canning, and livestock management marks significant growth. These skills enhance your self-reliance and contribute to a thriving homestead.

4. **Problem-Solving**: Overcoming challenges, whether it be pest infestations, crop failures, or resource shortages, showcases your resilience and adaptability. Each problem solved strengthens your knowledge and capabilities.

## Reflecting on Your Progress

Taking time to reflect on your progress allows you to appreciate how far you have come and identify areas for improvement. Reflection is a powerful tool for personal growth and enhancing your homesteading practices.

## Self-Assessment

1. **Strengths**: Identify the areas where you have excelled. This could be in producing abundant harvests, successfully raising livestock, or effectively managing resources. Acknowledge your strengths and build upon them.

2. **Challenges**: Recognize the challenges you have faced and how you addressed them. Consider what worked well and what could

be improved. Challenges offer valuable learning experiences that contribute to your overall growth.

3. **Achievements**: List your achievements, both big and small. From building infrastructure to developing new skills, every accomplishment is a step forward. Celebrating these milestones boosts motivation and confidence.

## Learning from Mistakes

Mistakes are inevitable on the journey to self-sufficiency. Instead of viewing them as setbacks, consider them opportunities for learning and improvement.

1. **Identify Mistakes**: Reflect on the mistakes you have made and their impact on your homesteading efforts. This could be anything from planting crops at the wrong time to mismanaging resources.

2. **Analyze and Learn**: Analyze why the mistake occurred and what you can learn from it. Consider alternative approaches and solutions that could prevent similar issues in the future.

3. **Adapt and Improve**: Use the insights gained from your mistakes to adapt and improve your practices. Continuous improvement is key to achieving long-term success in self-sufficiency.

## Planning for the Future

As you reflect on your progress, planning for the future becomes a crucial next step. Setting clear, achievable goals and developing a strategic plan will guide your efforts and ensure continued growth and success.

## Setting Future Goals

1. **Long-Term Vision**: Revisit your long-term vision for self-sufficiency. Consider what you want to achieve in the next five, ten, or twenty years. A clear vision provides direction and purpose.

2. **Specific Goals**: Break down your long-term vision into specific, measurable goals. These could include expanding your garden, increasing livestock, implementing new sustainable practices, or building additional infrastructure.

3. **Prioritizing Goals**: Prioritize your goals based on their importance and feasibility. Focus on the most critical and achievable goals first, then gradually work towards more ambitious objectives.

## Strategic Planning

1. **Action Plan**: Develop a detailed action plan for each goal. Outline the steps needed to achieve each objective, along with timelines and resources required. An action plan provides a clear roadmap for progress.

2. **Resource Allocation**: Assess the resources you have and those you need to achieve your goals. This includes financial resources, time, labor, and materials. Efficient resource allocation ensures that you can pursue your goals effectively.

3. **Risk Management**: Identify potential risks and challenges that could impact your plans. Develop strategies to mitigate these risks and have contingency plans in place. Being prepared for

uncertainties enhances resilience.

## Monitoring and Evaluation

1. **Progress Tracking**: Regularly track your progress towards your goals. Use journals, spreadsheets, or digital tools to record achievements, challenges, and lessons learned. Monitoring progress helps you stay on track and make informed decisions.

2. **Review and Adjust**: Periodically review your action plans and goals. Assess whether you are on track to achieve your objectives and make adjustments as needed. Flexibility and adaptability are essential for long-term success.

## Continuing Your Learning and Growth

The journey to self-sufficiency is an ongoing process that requires continuous learning and growth. Embracing lifelong learning ensures that you stay informed, innovative, and resilient in the face of new challenges.

## Expanding Knowledge

1. **Education and Training**: Pursue formal education and training opportunities related to homesteading, agriculture, and sustainability. This could include workshops, online courses, certifications, and degree programs.

2. **Books and Resources**: Continuously expand your knowledge by reading books, articles, and research papers. Stay updated on the latest advancements and best practices in your areas of interest.

3. **Networking and Mentorship**: Build relationships with experienced homesteaders, farmers, and experts. Participate in networking events, join homesteading communities, and seek mentorship. Learning from others' experiences and insights is invaluable.

## Experimentation and Innovation

1. **Trial and Error**: Embrace experimentation and be open to trying new methods and techniques. Not every experiment will succeed, but each one provides valuable insights and opportunities for innovation.

2. **Adopting New Technologies**: Stay informed about new technologies and tools that can enhance your homesteading practices. From advanced irrigation systems to sustainable energy solutions, integrating new technologies can improve efficiency and productivity.

3. **Creative Problem-Solving**: Foster a mindset of creative problem-solving. When faced with challenges, think outside the box and explore unconventional solutions. Innovation often arises from thinking differently and embracing new ideas.

## Community Engagement

1. **Sharing Knowledge**: Share your knowledge and experiences with others. This could be through teaching, writing, or participating in community events. Sharing knowledge not only helps others but also reinforces your own learning.

2. **Collaborative Projects**: Engage in collaborative projects with other homesteaders and community members. Working together on shared goals and initiatives strengthens community bonds and fosters collective growth.

3. **Advocacy and Outreach**: Advocate for sustainable practices and self-sufficiency within your community and beyond. Participate in outreach efforts to raise awareness and inspire others to pursue similar goals.

## Reflecting on the Bigger Picture

As you work towards self-sufficiency, it's important to reflect on the broader impact of your efforts. Consider how your actions contribute to environmental sustainability, community resilience, and global well-being.

1. **Environmental Stewardship**: Reflect on how your practices contribute to environmental health. Sustainable farming, renewable energy, and waste reduction efforts play a significant role in preserving natural resources and reducing ecological footprints.

2. **Community Impact**: Consider the positive impact you have on your local community. By sharing resources, knowledge, and support, you contribute to a stronger, more resilient community that can better withstand challenges and thrive together.

3. **Global Perspective**: Recognize that your efforts are part of a larger global movement towards sustainability and self-sufficiency. Each step you take contributes to a more sustainable future for everyone. Reflect on how your actions align with global goals, such as reducing carbon emissions, enhancing food security, and

promoting sustainable development.

Your journey to self-sufficiency is a dynamic and evolving process that requires dedication, reflection, and continuous learning. By celebrating your achievements, learning from your experiences, and planning for the future, you can continue to grow and thrive on your homestead. Embrace the challenges and opportunities that come your way, and remain committed to the principles of sustainability, resilience, and community.

As you move forward, remember that self-sufficiency is not just about achieving independence; it's about fostering a deeper connection to the land, nature, and the people around you. It's about creating a life that is rich in purpose, meaning, and fulfillment. Your journey is unique, and each step you take brings you closer to a more self-reliant and harmonious way of living.

Continue to seek knowledge, share your experiences, and inspire others to embark on their own journeys to self-sufficiency. Together, we can build a more sustainable and resilient world, one homestead at a time.

# Chapter 26

# Glossary of Terms

The world of homesteading and self-sufficiency involves a rich vocabulary that encompasses various disciplines, from agriculture to renewable energy. Understanding these terms is crucial for effectively communicating ideas and implementing practices. This glossary provides definitions and explanations of common terms and concepts that you may encounter on your journey to self-sufficiency.

## A

- **Acre**: A unit of land area used in the United States, equal to 43,560 square feet or about 4,047 square meters. One acre is often used to measure large plots of land for farming or homesteading.

- **Agriculture**: The practice of cultivating soil, growing crops, and raising animals for food, fiber, and other products used to sustain and enhance human life.

- **Anaerobic Digestion**: A biological process that breaks down organic matter in the absence of oxygen, producing biogas (methane and carbon dioxide) and digestate, a nutrient-rich byproduct used as fertilizer.

- **Annual**: A plant that completes its life cycle in one growing season, from germination to the production of seeds.

- **Aquaponics**: A system that combines aquaculture (raising fish) with hydroponics (growing plants in water) in a symbiotic environment. Fish waste provides nutrients for plants, and plants help filter and clean the water for fish.

# B

- **Backyard Chickens**: Raising chickens in a domestic backyard setting for eggs, meat, and companionship. This practice is popular in urban and suburban areas for its benefits in food production and pest control.

- **Biodiversity**: The variety of life in a particular habitat or ecosystem. High biodiversity is beneficial for ecosystem stability and resilience.

- **Biochar**: A form of charcoal produced from organic materials burned in a low-oxygen environment. Biochar is used as a soil amendment to improve fertility, water retention, and carbon sequestration.

- **Biodegradable**: Capable of being decomposed by bacteria or other biological means. Biodegradable materials break down naturally and reduce waste and pollution.

- **Biomass**: Organic material derived from plants and animals, used as a source of energy. Biomass can be burned directly for heat or converted into biofuels.

# C

- **Canning**: A method of preserving food by processing it in airtight containers. Canning prevents spoilage by killing bacteria, yeasts, and molds through heat processing.

- **Carbon Footprint**: The total amount of greenhouse gases produced directly and indirectly by human activities, expressed as carbon dioxide equivalent. Reducing carbon footprint involves minimizing energy use, waste, and emissions.

- **Chitin**: A natural polymer found in the exoskeletons of insects and crustaceans, and in fungal cell walls. Chitin is used in agriculture as a biopesticide and soil amendment to enhance plant growth and disease resistance.

- **Cold Frame**: A transparent-roofed enclosure, typically built low to the ground, used to protect plants from cold weather. Cold frames extend the growing season by providing a microclimate for seedlings and plants.

- **Companion Planting**: The practice of growing different plants together to benefit one or both of them. Companion plants can help with pest control, pollination, and nutrient uptake.

- **Compost**: Decomposed organic material used as a soil amendment. Compost improves soil structure, fertility, and microbial activity.

- **Cover Crop**: A crop grown primarily to improve soil health, manage soil erosion, and control pests and weeds. Cover crops are typically not harvested for food but are instead incorporated into the soil.

- **Crop Rotation**: The practice of growing different crops in succession on the same land to improve soil health, reduce pest and disease pressure, and increase biodiversity.

# D

- **Deep Water Culture (DWC)**: A hydroponic system where plants are suspended with their roots submerged in a nutrient-rich water solution. Oxygen is supplied to the roots using air stones or diffusers.

- **Drip Irrigation**: A method of watering plants using a network of tubes, pipes, and emitters that deliver water directly to the plant roots. Drip irrigation conserves water and reduces evaporation.

- **Drought-Resistant Plants**: Plants that can tolerate dry conditions and require minimal water once established. These plants are suitable for arid climates and water-scarce environments.

# E

- **Edible Landscape**: A landscaping approach that incorporates food-producing plants into ornamental and functional garden designs. Edible landscapes combine aesthetics with utility.

- **Energy Star**: A government-backed program that helps businesses and individuals save money and protect the environment through energy-efficient products and practices.

- **Erosion**: The process by which soil and rock are removed from the Earth's surface by wind, water, or other natural agents. Erosion can degrade land and reduce its fertility.

# F

- **Fermentation**: A metabolic process that converts sugar to acids, gases, or alcohol using microorganisms like yeast and bacteria. Fermentation is used in food preservation and production, such as in making yogurt, sauerkraut, and beer.

- **Food Forest**: A sustainable gardening method that mimics natural forests by incorporating diverse layers of plants, including trees, shrubs, herbs, groundcovers, and vines. Food forests provide food, habitat, and ecological benefits.

- **Food Sovereignty**: The right of people to define their own food systems and have access to healthy, culturally appropriate food produced through ecologically sound and sustainable methods.

- **Free-Range**: A farming method where animals are allowed to roam freely outdoors, rather than being confined in enclosures. Free-range systems promote animal welfare and natural behaviors.

# G

- **Germination**: The process by which a seed develops into a new plant. Germination requires the right conditions of moisture, temperature, and sometimes light.

- **Green Manure**: A cover crop that is grown specifically to be plowed under and incorporated into the soil, providing organic matter and nutrients to improve soil fertility and structure.

- **Greenhouse**: A structure with walls and roof made of transparent material, such as glass or plastic, used to grow plants in controlled environmental conditions. Greenhouses extend the growing season and protect plants from adverse weather.

- **Greywater**: Gently used water from sinks, showers, and washing machines. Greywater can be recycled for irrigation and other non-potable uses, reducing water consumption.

# H

- **Heirloom Variety**: A traditional plant variety that has been passed down through generations, often prized for its unique characteristics, flavor, and adaptability. Heirloom varieties are open-pollinated and genetically diverse.

- **Herbicide**: A chemical substance used to kill or inhibit the growth of unwanted plants (weeds). Herbicides can be selective (targeting specific plants) or non-selective (affecting all plants).

- **Homesteading**: A lifestyle of self-sufficiency that includes growing and preserving food, raising livestock, and making household items. Homesteading emphasizes sustainable and traditional practices.

- **Hydroponics**: A method of growing plants without soil, using nutrient-rich water. Hydroponic systems can be more efficient and productive than traditional soil-based gardening.

# I

- **Integrated Farm**: A farming system that combines crop production, livestock, aquaculture, and other agricultural activities in a way that maximizes efficiency and sustainability.

- **Integrated Pest Management (IPM)**: An ecological approach to pest control that combines biological, cultural, physical, and chemical methods. IPM aims to minimize pest damage while reducing reliance on pesticides.

- **Irrigation**: The artificial application of water to soil or land to assist in growing crops. Irrigation systems include sprinklers, drip lines, and flood irrigation.

# J

- **Jamming**: The process of making fruit preserves by cooking fruit with sugar and sometimes pectin. Jams are a popular method of preserving fruit.

# K

- **Kombucha**: A fermented tea beverage made with a symbiotic culture of bacteria and yeast (SCOBY). Kombucha is known for its probiotic content and health benefits.

# L

- **Legumes**: A family of plants that includes beans, peas, and lentils.

Legumes have nitrogen-fixing bacteria in their root nodules, which enrich the soil by converting atmospheric nitrogen into a form plants can use.

- **Livestock**: Domestic animals raised for agricultural purposes, such as meat, milk, eggs, wool, and labor. Common livestock includes cows, pigs, chickens, sheep, and goats.

# M

- **Mulch**: A layer of material, such as straw, wood chips, or leaves, applied to the surface of soil to conserve moisture, improve fertility, and reduce weed growth.

- **Mycelium**: The vegetative part of a fungus, consisting of a network of fine white filaments (hyphae). Mycelium plays a crucial role in decomposing organic matter and nutrient cycling.

- **Mycorrhiza**: A symbiotic association between fungi and plant roots. Mycorrhizal fungi enhance nutrient and water uptake for plants, while the fungi receive carbohydrates from the plant.

# N

- **Natural Farming**: A method of farming that relies on natural processes and materials, avoiding synthetic chemicals and genetically modified organisms (GMOs). It emphasizes soil health, biodiversity, and sustainability.

- **Nitrogen Fixation**: The process by which certain plants and microorganisms convert atmospheric nitrogen into a form that plants can use. Legumes and their symbiotic bacteria are well-known nitrogen fixers.

# O

- **Organic Farming**: A farming system that uses natural inputs and processes to grow crops and raise animals. Organic farming prohibits synthetic pesticides, herbicides, and fertilizers, and emphasizes soil health and biodiversity.

# P

- **Permaculture**: A design philosophy that creates sustainable and self-sufficient agricultural systems by mimicking natural ecosystems. Permaculture principles include diversity, resilience, and efficiency.

- **Polyculture**: The practice of growing multiple crop species in the same space to promote biodiversity and reduce pest and disease risks. Polyculture systems can improve soil health and yield stability.

- **Pollination**: The transfer of pollen from the male part of a flower (anther) to the female part (stigma), enabling fertilization and the production of seeds. Pollination is often facilitated by insects, birds, wind, and water.

- **Pollinator**: An organism that transfers pollen from one flower to another, facilitating plant reproduction. Pollinators include bees, butterflies, birds, and bats, and are essential for many food crops.

# Q

- **Quinoa**: A high-protein, gluten-free grain-like seed that is cultivated for its edible seeds. Quinoa is known for its nutritional value and versatility in cooking.

# R

- **Rainwater Harvesting**: The practice of collecting and storing rainwater for later use. Rainwater harvesting systems can include barrels, cisterns, and ponds, and the water can be used for irrigation, livestock, and household needs.

- **Raised Bed**: A gardening method where soil is enclosed in a frame above ground level. Raised beds improve drainage, reduce soil compaction, and can be filled with high-quality soil for better plant growth.

- **Renewable Energy**: Energy from sources that are naturally replenishing, such as solar, wind, hydro, and biomass. Renewable energy systems reduce reliance on fossil fuels and decrease environmental impact.

# S

- **Seed Saving**: The practice of collecting and storing seeds from plants for future planting. Seed saving helps preserve heirloom varieties, maintain genetic diversity, and reduce dependency on commercial seed sources.

- **Soil Amendment**: A material added to soil to improve its physical properties, fertility, and biological activity. Common soil amendments include compost, manure, lime, and biochar.

- **Sustainable Agriculture**: Farming practices that meet current food needs without compromising the ability of future generations to meet theirs. Sustainable agriculture emphasizes environmental health, economic viability, and social equity.

# T

- **Thermal Mass**: A material that absorbs, stores, and releases heat energy, helping to regulate temperature fluctuations. Thermal mass is used in passive solar design to enhance energy efficiency.

- **Tillage**: The preparation of soil for planting by mechanical agitation, such as plowing, turning, or stirring. While tillage can improve soil aeration and weed control, excessive tillage can lead to soil degradation and erosion.

- **Transplanting**: The process of moving a plant from one location to another. Transplants are typically started indoors or in a greenhouse and then moved to the garden when they are strong enough to survive.

# U

- **Urban Farming**: The practice of cultivating, processing, and distributing food in or around urban areas. Urban farming includes community gardens, rooftop gardens, and vertical farming.

# V

- **Vermiculture**: The cultivation of worms for composting organic waste. Worms, particularly red wigglers, break down organic matter into nutrient-rich vermicompost, which can be used as a soil amendment.

- **Vermicompost**: The product of the decomposition of organic material by earthworms. Vermicompost is rich in nutrients and beneficial microorganisms, making it an excellent soil amendment.

# W

- **Water Conservation**: Practices and techniques aimed at reducing water use and preserving water resources. Water conservation methods include efficient irrigation, rainwater harvesting, and xeriscaping.

- **Water Table**: The level below which the ground is saturated with water. The water table fluctuates based on precipitation, irrigation, and groundwater extraction.

- **Windbreak**: A barrier, such as trees or shrubs, planted to reduce wind speed and protect crops, soil, and livestock from wind damage. Windbreaks also provide habitat for wildlife and improve biodiversity.

# X

- **Xeriscaping**: A landscaping method that reduces or eliminates the need for irrigation by using drought-tolerant plants and efficient water use techniques. Xeriscaping is particularly useful in arid regions.

# Y

- **Yurt**: A portable, round tent traditionally used by nomadic peoples in Central Asia. Modern yurts are often used as sustainable living spaces, offering a simple and efficient housing option.

# Z

- **Zero Waste**: A philosophy and lifestyle that aims to minimize waste generation by reusing, recycling, composting, and reducing consumption. Zero waste practices promote sustainability and environmental responsibility.

# Chapter 27

# Seasonal Planting Guides

Seasonal planting guides are essential tools for every homesteader, ensuring that crops are planted at the optimal times for growth, yield, and resilience. Understanding the best times to plant various crops allows you to maximize your harvest, maintain soil health, and extend the growing season. This appendix provides detailed seasonal planting guides tailored to different climates and regions, along with practical tips and considerations for each season.

## Understanding Planting Zones

Before delving into specific planting schedules, it's crucial to understand USDA Plant Hardiness Zones, which help determine the best planting times for various crops based on average minimum winter temperatures. Knowing your planting zone will guide your decisions on when to start seeds, transplant seedlings, and direct sow crops.

### How to Determine Your Planting Zone

1. **USDA Hardiness Zone Map**: Refer to the USDA Plant Hardiness Zone Map, which divides North America into 13 zones based on average annual minimum winter temperatures. Each zone is further divided into subzones (e.g., 5a, 5b).

2. **Local Extension Offices**: Consult your local agricultural extension office for specific information about your region's climate and planting recommendations.

3. **Online Tools**: Use online tools and apps that can pinpoint your location and provide detailed planting zone information.

# Spring Planting Guide

Spring is a time of renewal and growth, making it an ideal season for planting many types of crops. As temperatures warm and days lengthen, both cool-season and warm-season crops can be planted, either directly in the ground or as transplants.

## Cool-Season Crops

Cool-season crops thrive in the mild temperatures of early spring. These crops can tolerate light frosts and prefer growing conditions that are too cold for most warm-season vegetables.

## Early Spring (March - April)

1. **Peas**: Sow peas as soon as the soil can be worked. Varieties include shelling peas, snap peas, and snow peas. Plant seeds 1 inch deep and 2 inches apart in rows.

2. **Spinach**: Sow spinach seeds directly in the garden, 1/2 inch deep and 2 inches apart. Spinach prefers well-drained soil and full sun.

3. **Radishes**: Radishes are fast-growing and can be planted in early spring. Sow seeds 1/2 inch deep and 1 inch apart. Thin seedlings to 2 inches apart.

4. **Lettuce**: Lettuce can be direct-seeded or transplanted. Sow seeds 1/4 inch deep and 1 inch apart. Thin to 6-12 inches apart, depending on the variety.

5. **Carrots**: Sow carrot seeds 1/4 inch deep and 2 inches apart. Carrots prefer loose, well-drained soil. Thin seedlings to 3 inches apart.

## Mid to Late Spring (April - May)

1. **Broccoli**: Start broccoli seeds indoors 6-8 weeks before the last frost date. Transplant seedlings outdoors when they are 4-6 weeks old and the danger of frost has passed.

2. **Cabbage**: Start cabbage seeds indoors 6-8 weeks before the last frost. Transplant seedlings outdoors when they are 4-6 weeks old.

3. **Cauliflower**: Like broccoli and cabbage, start cauliflower seeds indoors and transplant outdoors. Provide consistent moisture for best growth.

4. **Beets**: Direct sow beet seeds 1/2 inch deep and 2 inches apart. Thin seedlings to 4 inches apart. Beets prefer cool temperatures and well-drained soil.

5. **Potatoes**: Plant seed potatoes 4 inches deep and 12 inches apart in rows. Potatoes prefer cool weather and well-drained soil.

## Warm-Season Crops

Warm-season crops are sensitive to frost and should be planted after the danger of frost has passed. These crops require warmer soil and air temperatures to thrive.

## Late Spring (May - June)

1. **Tomatoes**: Start tomato seeds indoors 6-8 weeks before the last frost date. Transplant seedlings outdoors after the danger of frost has passed and the soil has warmed.

2. **Peppers**: Start pepper seeds indoors 8-10 weeks before the last frost. Transplant seedlings outdoors after the soil has warmed and nighttime temperatures are consistently above 50°F.

3. **Cucumbers**: Direct sow cucumber seeds 1 inch deep and 6 inches apart in rows. Alternatively, start seeds indoors 3-4 weeks before transplanting outdoors.

4. **Squash (Summer and Winter)**: Direct sow squash seeds 1 inch deep and 3 feet apart in rows. For transplants, start seeds indoors 3-4 weeks before the last frost.

5. **Beans (Bush and Pole)**: Direct sow bean seeds 1 inch deep and 2-3 inches apart in rows. Bush beans mature faster, while pole beans require support and produce over a longer period.

# Summer Planting Guide

Summer is the peak growing season for many warm-season crops. As temperatures rise and days lengthen, gardens can produce abundant harvests. Summer planting also includes succession planting to ensure continuous yields.

## Warm-Season Crops

Warm-season crops planted in late spring continue to thrive and produce throughout the summer. Additional plantings can extend the harvest period.

## Early Summer (June - July)

1. **Corn**: Direct sow corn seeds 1 inch deep and 8-12 inches apart in rows. Plant multiple rows for better pollination. Corn prefers full sun and fertile soil.

2. **Melons (Watermelon, Cantaloupe)**: Direct sow melon seeds 1 inch deep and 18-24 inches apart in rows. Melons require warm soil and full sun to thrive.

3. **Eggplant**: Start eggplant seeds indoors 8-10 weeks before the last frost. Transplant seedlings outdoors after the soil has warmed and nighttime temperatures are consistently above 50°F.

4. **Okra**: Direct sow okra seeds 1 inch deep and 12-18 inches apart in rows. Okra thrives in hot weather and requires full sun.

5. **Basil**: Direct sow basil seeds or transplant seedlings. Basil prefers

warm temperatures, full sun, and well-drained soil.

## Mid to Late Summer (July - August)

1. **Bush Beans**: For a continuous harvest, sow bush beans every 2-3 weeks until mid-summer. Beans prefer warm temperatures and full sun.

2. **Zucchini**: Direct sow zucchini seeds or transplant seedlings. Zucchini grows quickly and prefers warm weather and full sun.

3. **Swiss Chard**: Direct sow Swiss chard seeds 1/2 inch deep and 2-3 inches apart in rows. Chard tolerates heat and can be harvested continuously.

4. **Collard Greens**: Direct sow collard seeds or transplant seedlings. Collards tolerate heat but prefer cooler temperatures for optimal growth.

5. **Cilantro**: Direct sow cilantro seeds every 2-3 weeks for a continuous harvest. Cilantro prefers cooler temperatures but can be grown in partial shade during the summer.

## Fall Planting Guide

Fall is a time for harvesting and preparing for the next growing season. Many cool-season crops thrive in the cooler temperatures of fall, and some warm-season crops can be planted for a late harvest.

## Cool-Season Crops

Cool-season crops planted in late summer and early fall can extend the harvest season and provide fresh produce into the cooler months.

## Late Summer (August - September)

1. **Broccoli**: Start broccoli seeds indoors 6-8 weeks before the first frost date. Transplant seedlings outdoors in late summer for a fall harvest.

2. **Brussels Sprouts**: Start Brussels sprouts seeds indoors and transplant seedlings outdoors. Brussels sprouts require a long growing season and cooler temperatures to mature.

3. **Cabbage**: Start cabbage seeds indoors and transplant seedlings outdoors. Cabbage grows well in cool weather and can be harvested into late fall.

4. **Carrots**: Direct sow carrot seeds 1/4 inch deep and 2 inches apart. Carrots can be harvested throughout the fall and stored for winter use.

5. **Kale**: Direct sow kale seeds or transplant seedlings. Kale tolerates frost and can be harvested throughout the fall and winter.

## Early Fall (September - October)

1. **Garlic**: Plant garlic cloves in the fall, 1-2 inches deep and 4-6 inches apart. Garlic requires a cold period to develop properly and will be ready to harvest the following summer.

2. **Spinach**: Direct sow spinach seeds 1/2 inch deep and 2 inches apart. Spinach prefers cool temperatures and can be harvested into early

winter.

3. **Radishes**: Direct sow radish seeds 1/2 inch deep and 1 inch apart. Radishes mature quickly and can be harvested in as little as 3-4 weeks.

4. **Turnips**: Direct sow turnip seeds 1/2 inch deep and 2-3 inches apart. Turnips grow well in cool weather and can be harvested into late fall.

5. **Beets**: Direct sow beet seeds 1/2 inch deep and 2 inches apart. Beets tolerate light frost and can be harvested into late fall.

## Winter Planting Guide

Winter planting is less common in colder climates, but in milder regions or with the use of season extension techniques, you can grow certain crops throughout the winter.

## Cold-Season Crops

Cold-season crops are hardy and can withstand colder temperatures, frost, and even light snow.

## Late Fall (October - November)

1. **Onions (Sets and Seed)**: Plant onion sets or seeds in late fall for an early spring harvest. Onions prefer well-drained soil and full sun.

2. **Leeks**: Start leek seeds indoors and transplant seedlings outdoors in late fall. Leeks tolerate cold weather and can be harvested

throughout the winter.

3. **Winter Lettuce**: Direct sow winter lettuce varieties or transplant seedlings. Winter lettuce can be grown under row covers or in cold frames for protection.

4. **Mâche (Corn Salad)**: Direct sow mâche seeds in late fall. Mâche is very cold-hardy and can be harvested throughout the winter.

5. **Parsley**: Direct sow parsley seeds or transplant seedlings. Parsley is frost-tolerant and can be harvested throughout the winter.

## Overwintering Techniques

To grow crops successfully in the winter, use season extension techniques to protect plants from harsh weather conditions.

1. **Cold Frames**: Build cold frames to protect plants from frost and extend the growing season. Cold frames can be made from wood, metal, or plastic and covered with glass or clear plastic.

2. **Row Covers**: Use floating row covers to protect plants from frost and cold winds. Row covers can be made from lightweight fabric and are draped directly over plants.

3. **Mulching**: Apply a thick layer of mulch around plants to insulate the soil and protect roots from freezing. Straw, leaves, and wood chips are effective mulching materials.

4. **Greenhouses**: Utilize greenhouses to create a controlled environment for growing crops in the winter. Greenhouses provide warmth, light, and protection from adverse weather.

5. **Hoop Houses**: Construct hoop houses using PVC pipes or metal hoops covered with clear plastic. Hoop houses provide a microclimate for growing cold-season crops in the winter.

## Practical Tips for Seasonal Planting

### Soil Preparation

1. **Testing Soil**: Before planting, test your soil to determine its pH, nutrient levels, and texture. Amend the soil based on test results to ensure optimal growing conditions.

2. **Adding Compost**: Incorporate compost into the soil to improve fertility, structure, and microbial activity. Compost provides essential nutrients and enhances soil health.

3. **Mulching**: Apply mulch to garden beds to conserve moisture, suppress weeds, and regulate soil temperature. Organic mulches like straw, wood chips, and leaves are beneficial.

### Watering and Irrigation

1. **Consistent Watering**: Provide consistent watering to keep the soil evenly moist but not waterlogged. Avoid overwatering, which can lead to root rot and other issues.

2. **Drip Irrigation**: Use drip irrigation systems to deliver water directly to plant roots, reducing evaporation and ensuring efficient water use.

3. **Mulching**: Mulch around plants to retain soil moisture and reduce the need for frequent watering. Mulch also helps regulate soil temperature.

## Pest and Disease Management

1. **Crop Rotation**: Practice crop rotation to reduce pest and disease pressure. Rotating crops disrupts pest and disease cycles and improves soil health.

2. **Companion Planting**: Use companion planting to deter pests and enhance plant growth. Certain plants can repel pests, attract beneficial insects, or improve soil fertility.

3. **Integrated Pest Management (IPM)**: Implement IPM strategies to manage pests and diseases. IPM combines biological, cultural, physical, and chemical methods to minimize pest damage.

## Season Extension

1. **Succession Planting**: Practice succession planting to ensure a continuous harvest. Plant crops in intervals to maintain a steady supply of fresh produce.

2. **Using Cold Frames and Greenhouses**: Utilize cold frames and greenhouses to extend the growing season and protect plants from adverse weather.

3. **Row Covers and Hoop Houses**: Use row covers and hoop houses to create microclimates for growing crops in cooler weather. These structures protect plants from frost and cold winds.

# Record Keeping

1. **Garden Journal**: Keep a garden journal to record planting dates, weather conditions, pest and disease occurrences, and harvest yields. A journal helps track progress and plan for future seasons.

2. **Planting Calendar**: Create a planting calendar based on your region's climate and planting zones. A calendar helps organize planting schedules and ensures timely planting.

3. **Photos and Diagrams**: Take photos and create diagrams of your garden layout each season. Visual records help you remember what worked well and what needs improvement.

# Chapter 28

# Troubleshooting Common Problems

Maintaining a homestead is a rewarding but complex task that involves constant vigilance and problem-solving. This appendix provides a comprehensive guide to troubleshooting common problems you might encounter. By understanding these issues and their solutions, you can ensure the health and productivity of your homestead.

## Plant Health Issues

### Nutrient Deficiencies

Plants need a variety of nutrients to grow properly. Identifying and correcting nutrient deficiencies is crucial for maintaining healthy crops.

### Nitrogen Deficiency

**Symptoms**: Yellowing leaves, stunted growth, poor fruit production.

**Solutions:**

- **Compost and Manure:** Add compost or well-rotted manure to the soil to increase nitrogen levels.

- **Green Manures:** Plant nitrogen-fixing cover crops such as clover or beans.

- **Organic Fertilizers:** Apply blood meal, fish emulsion, or other organic nitrogen sources.

## Phosphorus Deficiency

**Symptoms:** Dark green or purplish leaves, slow growth, weak root systems.

**Solutions:**

- **Bone Meal:** Incorporate bone meal into the soil to provide phosphorus.

- **Rock Phosphate:** Use rock phosphate as a long-term phosphorus source.

- **Balanced Fertilizer:** Apply a balanced organic fertilizer containing phosphorus.

## Potassium Deficiency

**Symptoms:** Yellowing or browning leaf edges, weak stems, poor fruit and flower development.

**Solutions:**

- **Potash**: Add potash or wood ash to the soil to increase potassium levels.

- **Compost**: Use compost that includes potassium-rich materials like banana peels.

- **Seaweed Extracts**: Apply liquid seaweed extracts to provide a quick potassium boost.

## Pests and Diseases

Pests and diseases can devastate your crops if not managed properly. Integrated Pest Management (IPM) strategies are essential for controlling these issues.

### Common Pests

#### Aphids

**Symptoms**: Curling, yellowing leaves, sticky honeydew on leaves, presence of sooty mold.

**Solutions**:

- **Beneficial Insects**: Introduce ladybugs, lacewings, and parasitic wasps.

- **Neem Oil**: Spray neem oil to deter aphids.

- **Insecticidal Soap**: Use insecticidal soap to kill aphids on contact.

#### Caterpillars

**Symptoms**: Holes in leaves, defoliation, presence of frass (caterpillar droppings).

**Solutions**:

- **Handpicking**: Remove caterpillars by hand.

- **Bacillus thuringiensis (Bt)**: Apply Bt, a natural bacterial pesticide.

- **Row Covers**: Use row covers to prevent butterflies and moths from laying eggs on plants.

### Slugs and Snails

**Symptoms**: Irregular holes in leaves, slime trails on plants and soil.

**Solutions**:

- **Beer Traps**: Place shallow dishes of beer to attract and drown slugs.

- **Copper Barriers**: Use copper tape or wire around plants to repel slugs.

- **Handpicking**: Collect and remove slugs and snails at night.

## Common Diseases

### Powdery Mildew

**Symptoms**: White, powdery spots on leaves, stems, and flowers.

**Solutions**:

- **Remove Affected Parts**: Prune and dispose of infected plant parts.

- **Neem Oil**: Spray neem oil to prevent the spread of mildew.

- **Baking Soda Solution**: Apply a solution of baking soda and water to affected plants.

**Blight (Early and Late)**

**Symptoms**: Dark, water-soaked spots on leaves and stems, rapidly spreading lesions.

**Solutions**:

- **Crop Rotation**: Rotate crops to prevent the buildup of blight pathogens.

- **Resistant Varieties**: Plant blight-resistant varieties.

- **Fungicides**: Use organic fungicides like copper or sulfur-based products.

**Root Rot**

**Symptoms**: Wilting, yellowing leaves, blackened roots, and stems.

**Solutions**:

- **Improve Drainage**: Ensure proper soil drainage to prevent waterlogging.

- **Fungicides**: Apply organic fungicides to affected plants.

- **Avoid Overwatering**: Water plants only when necessary to prevent root rot.

# Soil Health Issues

## Soil Compaction

**Symptoms**: Poor water infiltration, stunted root growth, reduced crop yields.

**Solutions**:

- **Aeration**: Use a garden fork or aerator to loosen compacted soil.

- **Organic Matter**: Incorporate organic matter such as compost or cover crops to improve soil structure.

- **Avoid Heavy Machinery**: Minimize the use of heavy equipment that can compact soil.

## Soil Erosion

**Symptoms**: Loss of topsoil, exposed roots, poor plant growth.

**Solutions**:

- **Mulching**: Apply mulch to protect soil from erosion.

- **Cover Crops**: Plant cover crops to hold soil in place.

- **Terracing**: Create terraces on slopes to reduce runoff and erosion.

## Soil pH Imbalance

**Symptoms**: Yellowing leaves, poor growth, nutrient deficiencies.

**Solutions**:

- **Soil Testing**: Test soil pH to determine acidity or alkalinity.

- **Lime**: Add lime to raise soil pH in acidic soils.

- **Sulfur**: Apply sulfur to lower soil pH in alkaline soils.

## Water Management Issues

### Overwatering

**Symptoms**: Yellowing leaves, root rot, mold growth, and reduced plant vigor.

**Solutions**:

- **Check Soil Moisture**: Test soil moisture before watering to ensure plants need it.

- **Improve Drainage**: Amend soil to improve drainage and prevent waterlogging.

- **Watering Schedule**: Establish a consistent watering schedule that meets the needs of your plants without overwatering.

### Underwatering

**Symptoms**: Wilting, dry leaves, stunted growth, and reduced yields.

**Solutions**:

- **Mulching**: Apply mulch to retain soil moisture.

- **Drip Irrigation**: Install drip irrigation to provide consistent moisture to plants.

- **Watering Routine**: Water plants deeply and less frequently to encourage deep root growth.

## Water Quality

**Symptoms**: Poor plant growth, leaf discoloration, salt buildup.

**Solutions**:

- **Test Water**: Test irrigation water for contaminants and pH levels.

- **Filtration Systems**: Use filtration systems to remove impurities from water.

- **Rainwater Harvesting**: Collect and use rainwater for irrigation to reduce dependence on treated water sources.

# Livestock Health Issues

## Common Livestock Health Problems

### Parasites

**Symptoms**: Weight loss, poor coat condition, diarrhea, and anemia.

**Solutions**:

- **Regular Deworming**: Implement a regular deworming schedule using natural or conventional dewormers.

- **Pasture Management**: Rotate pastures to break parasite life cycles.

- **Herbal Remedies**: Use herbal remedies like garlic and diatomaceous earth to control parasites.

## Respiratory Infections

**Symptoms**: Coughing, nasal discharge, labored breathing.

**Solutions**:

- **Proper Ventilation**: Ensure barns and coops have adequate ventilation to reduce humidity and ammonia buildup.

- **Clean Living Spaces**: Keep livestock areas clean and dry to prevent infections.

- **Isolate Infected Animals**: Isolate and treat infected animals to prevent the spread of disease.

## Foot Rot

**Symptoms**: Lameness, swelling, foul-smelling discharge from hooves.

**Solutions**:

- **Foot Baths**: Use foot baths with zinc sulfate or copper sulfate to treat and prevent foot rot.

- **Regular Trimming**: Trim hooves regularly to prevent overgrowth and infection.

- **Dry Bedding**: Maintain dry bedding to reduce the risk of foot rot.

## Pest Control Issues

### Integrated Pest Management (IPM)

**Steps to Implement IPM**:

1. **Identify Pests**: Accurately identify pests to choose appropriate control methods.

2. **Monitor Pests**: Regularly monitor pest populations and damage levels.

3. **Set Action Thresholds**: Determine the pest population levels at which action should be taken.

4. **Implement Controls**: Use a combination of biological, cultural, mechanical, and chemical controls.

5. **Evaluate Effectiveness**: Assess the effectiveness of pest control methods and adjust as needed.

### Common Garden Pests and Solutions

### Japanese Beetles

**Symptoms**: Skeletonized leaves, damaged flowers and fruits.

**Solutions**:

- **Handpicking**: Remove beetles by hand and dispose of them.

- **Neem Oil**: Spray neem oil to deter beetles.

- **Milky Spore**: Apply milky spore powder to kill beetle grubs in the soil.

## Squash Bugs

**Symptoms**: Wilting plants, yellowing leaves, and presence of egg masses.

**Solutions**:

- **Handpicking**: Remove bugs and egg masses by hand.

- **Diatomaceous Earth**: Apply diatomaceous earth around plants to deter bugs.

- **Trap Crops**: Plant trap crops to attract squash bugs away from main crops.

## Tomato Hornworms

**Symptoms**: Large holes in leaves, defoliation, presence of large green caterpillars.

**Solutions**:

- **Handpicking**: Remove hornworms by hand.

- **Beneficial Insects**: Introduce parasitic wasps that prey on hornworms.

- **Bt (Bacillus thuringiensis)**: Apply Bt to control hornworm populations.

## Weather-Related Issues

### Frost Damage

**Symptoms**: Blackened, wilted leaves, stunted growth.

**Solutions**:

- **Frost Covers**: Use frost covers or blankets to protect plants.

- **Watering**: Water plants before a frost to help retain heat in the soil.

- **Cold Frames**: Utilize cold frames to extend the growing season and protect against frost.

### Heat Stress

**Symptoms**: Wilting, yellowing leaves, sunscald, poor fruit set.

**Solutions**:

- **Mulching**: Apply mulch to keep soil cool and retain moisture.

- **Shade Cloth**: Use shade cloth to protect plants from intense sun.

- **Consistent Watering**: Ensure plants receive adequate water during heat waves.

## Wind Damage

**Symptoms**: Broken stems, uprooted plants, leaf burn.

**Solutions**:

- **Windbreaks**: Plant trees or shrubs as windbreaks to protect gardens.

- **Staking**: Stake plants to provide support and prevent wind damage.

- **Row Covers**: Use row covers to shield plants from strong winds.

## Harvest and Storage Issues

### Poor Yield

**Symptoms**: Low fruit or vegetable production, small or misshapen produce.

**Solutions**:

- **Soil Testing**: Test soil and amend with necessary nutrients.

- **Proper Spacing**: Ensure plants are spaced correctly to avoid competition for resources.

- **Pollination**: Encourage pollinators by planting flowers and providing habitat.

### Post-Harvest Storage

**Symptoms**: Spoilage, mold growth, loss of flavor or texture.

**Solutions**:

- **Proper Harvesting**: Harvest crops at the right maturity stage and handle gently to avoid damage.

- **Storage Conditions**: Store produce in optimal conditions (cool, dark, and dry) to prolong shelf life.

- **Preservation Methods**: Use canning, freezing, drying, or fermenting to preserve excess produce.

## General Maintenance Issues

### Equipment Maintenance

**Symptoms**: Poor performance, breakdowns, increased fuel consumption.

**Solutions**:

- **Regular Servicing**: Follow manufacturer guidelines for regular maintenance and servicing of equipment.

- **Proper Storage**: Store equipment in a dry, sheltered location to prevent rust and deterioration.

- **Sharpening Tools**: Keep blades and cutting tools sharp to ensure efficient operation.

### Infrastructure Maintenance

**Symptoms**: Structural damage, leaks, inefficient systems.

**Solutions**:

- **Routine Inspections**: Conduct regular inspections of buildings, fences, and irrigation systems.

- **Timely Repairs**: Address repairs promptly to prevent further damage.

- **Upgrades**: Invest in upgrades to improve efficiency and durability.

# Human and Community Issues

## Burnout

**Symptoms**: Fatigue, loss of motivation, decreased productivity.

**Solutions**:

- **Time Management**: Prioritize tasks and delegate responsibilities.

- **Self-Care**: Take regular breaks, maintain a healthy diet, and get adequate rest.

- **Support Networks**: Build a support network of family, friends, and fellow homesteaders.

## Community Relations

**Symptoms**: Conflicts, lack of support, isolation.

**Solutions**:

- **Communication**: Foster open and respectful communication with neighbors and community members.

- **Involvement**: Participate in local events, farmers' markets, and community groups.

- **Collaboration**: Collaborate on projects and share resources to build strong community ties.

## Learning and Adaptation

**Symptoms**: Stagnation, resistance to change, lack of knowledge.

**Solutions**:

- **Continuous Education**: Attend workshops, read books, and participate in online courses to stay informed.

- **Experimentation**: Try new methods and techniques to improve your homesteading practices.

- **Adaptability**: Stay flexible and willing to adapt to changing conditions and new information.